NEW ZEALAND
IN
FLOWER

ANDRIS APSE LTD

NEW ZEALAND IN FLOWER

An Illustrated Guide to Native Flowering Plants

Text
Alison Evans

Picture editing
Stephen Barnett

RM
REED METHUEN

ACKNOWLEDGEMENTS

The photographs which appear in this book are the work of many of New Zealand's finest wildlife photographers (see page vi) and I wish to acknowledge their encouragement and patience during the long period of selection and production that a book of this kind entails. Special thanks go to Rod Morris.

S.B.

For encouragement, proof-reading and many helpful comments on the text: Bram Evans, Peter Johnson, Stephen Barnett, Rod Morris, Kath Dickinson, Geoff Baylis, Brent McKenzie and Alan Bennington.

For identifications and advice on nomenclature: Alan Mark, Phil Garnock-Jones, Eric Godley, Tony Druce and Ian St George.

A.E.

A Bookmakers Book

First published 1987 by Reed Methuen Publishers Ltd
Private Bag, Birkenhead, Auckland

ISBN 0 474 00263 2

Design and production by Stephen Barnett/Bookmakers, Auckland
Typesetting by Edsetera and Typocrafters, Auckland
Printed by Kyodo-Shing Loong Printing, Singapore

Front cover photographs show (top) *Myosotidium hortensia* (photograph by Brett Mackay), and (bottom left) *Clianthus puniceus*, (bottom centre) *Thelymitra hatchii*, (bottom right) *Sophora microphylla* (all by Rod Morris). Back cover photograph shows *Ranunculus insignis* (E. & J. Kelly). Title page photograph is by Andris Apse.

CONTENTS

THE PHOTOGRAPHERS

The authors are indebted to the following photographers for their encouragement and assistance. Special mention must be made of Jack Bedford whose death occurred during production of the book. His contribution to the illustration of *New Zealand in Flower* was both important and great in the number of photographs reproduced.

Photographers are credited alongside their photographs with the initials shown here:

A.E.— **Alison Evans**
C.E.B.— **C. E. Barwell**
J.B.— **Jack Bedford**
P.E.C.— **Peter E. Cameron**
B.E.— **Brian Enting**
J.F.— **John Fennell**
B.G.— **Brenda Given**
D.G.— **David Given**
D.H.— **Don Hadden**
P.N.J.— **Peter Johnson**
E.&J.K.— **Errol & Jan Kelly**
A.F.M.— **Alan F. Mark**
R.M.— **Rod Morris**
A.&S.P.— **Andrew & Sue Penniket**
C.P.— **Craig Potton**
G.R.— **Gordon Roberts**
G.R.R.— **G. R. (Dick) Roberts**
P.M.&J.L.S.— **Paul M. & Joy L. Sagar**
M.F.S.— **M. F. Soper**
P.S.— **P. Syms**
G.W.— **Gordon Watson**
R.W.— **Ron Willems**

The photograph on page i is by Brett Mackay; that across pages ii and iii is by Andris Apse; and the contents page photograph (p. v) is by Rod Morris.

AUTHORS' NOTE

New Zealand in Flower presents an illustrated selection of this region's most beautiful, conspicuous and interesting indigenous flowering plants grouped according to the habitats in which they most often occur. The choice of broad habitat categories and the designation of species within them inevitably involved making generalisations, such as the inclusion of lowland grasslands in the section on Scrub, Shrubland, Open Places, and upland grasslands in Uplands and Mountains. Species occurring in more than one habitat are cross-referenced from lists included in the habitat introductions.

The family name of each plant appears in capital letters at the end of the relevant subject text. Abbreviations used in the texts are **sp.** (species — singular), **spp.** (species — plural), **ssp.** (subspecies) and **var.** (variety). The index lists the botanical name and any common name of all plant subjects.

For the purposes of this book, the New Zealand region is taken to include Macquarie Island. Although Macquarie Island is, in fact, an Australian territory, its flora is related to the New Zealand region.

The spelling of a few species' names may appear unfamiliar but conforms to very recent official changes.

INTRODUCTION

Alison Evans

For its size, latitude and topographical diversity, New Zealand has a relatively small flora of about 2000 species of higher plants — ferns and fern allies, gymnosperms and flowering plants. Yet it is a flora like no other, as Joseph Banks, one of the first botanists to see and describe New Zealand's plants, observed: 'The intire novelty however of the greatest part of what we found recompens'd us as natural historians for the want of variety.' The novelty of which Banks wrote is indeed a special feature of our native plants. Thus of some 1800 species of indigenous flowering plants, 84 per cent are endemic, that is, they are confined to the New Zealand botanical region.

Although the origins of different genera continue to be debated by botanists and biogeographers, it is generally agreed that New Zealand's plants derive from two main groups. First are those of Gondwanaland, the ancient 'supercontinent' from which New Zealand separated some 80 million years ago. Some of our plants evolved before this separation: these include gymnosperms such as *Podocarpus* (totara), *Phyllocladus* (celery pines), *Agathis* (kauri), *Dacrycarpus* (kahikatea) and flowering plants such as *Pseudowintera* (pepper trees), *Nothofagus* (southern beeches), *Griselinia* (broadleaf) and *Metrosideros* (ratas, pohutukawa).

However, Gondwanaland plants that evolved later, after New Zealand broke adrift, would have had to cross the Southern Ocean to reach here. To do this they must have had either fleshy, bird-dispersed fruits such as in *Astelia*, *Pseudopanax* and *Fuchsia*, or very light seeds, as in *Gentiana*.

The same methods of transport would have applied to the other main group from which New Zealand's flora is derived — those migrants carried across the Tasman Sea in the direction of the prevailing westerlies. Likely arrivals in this category include hebes and orchids (which have very light seeds), *Celmisia* and *Parsonsia* (whose seeds are specialised for travelling on wind currents), and *Toronia*, *Vitex* and *Carpodetus* (whose fleshy fruits could have been brought by birds).

The westerly winds that have always been important in carrying new species to New Zealand continue to transport plants and birds today. Some relatively recent arrivals include the willowherb *Epilobium gunnianum* in the north-western South Island, the tongue orchid *Cryptostylis subulata* in North Auckland, and *Sprengelia incarnata* in south-west Fiordland. Others very rare in New Zealand but common in Australia, such as *Myoporum debile*, may well be recent trans-Tasman waifs. The entry of a number of bird species into New Zealand has also been assisted by the westerlies. Birds thought to have arrived in this way in relatively recent times include, for example, the silvereye (*Zosterops lateralis lateralis*), spur-winged plover (*Vanellus miles novaehollandiae*), welcome swallow (*Hirundo tahitica neoxena*), white-faced heron (*Ardea novaehollandiae novaehollandiae*) and Australian coot (*Fulica atra australis*).

Most of the genera in both main groups occur in other southern lands — only 35 (or 11 per cent) of our 314 indigenous flowering plant genera are endemic (compared with 84 per cent endemic species), and most of these contain just one (e.g. *Corallospartium*, *Myosotidium*, *Hectorella*) or a few species (e.g. *Hoheria*, *Stilbocarpa*, *Notothlaspi*).

We share some 50 genera and 200 species with Australia. In some (e.g. *Swainsona*, *Clianthus*, *Toronia*), the single endemic New Zealand species is an outlier of an essentially Australian genus, while in others, particularly alpine genera (e.g. *Hebe*, *Aciphylla*, *Celmisia*), the reverse applies — Australia has just one or a few species and the vast majority occur in New Zealand, where they have presumably diversified to a greater extent than where the group originated.

New Zealand has some 23 genera and 60 species in common with South America, although many of these occur in south-east Australia, Tasmania and some Pacific islands as well. Included here are *Sophora*, *Hebe*, *Aristotelia*, *Griselinia*, *Donatia*, *Uncinia*, *Acaena*, *Ourisia* and species such as *Sophora microphylla*, *Hebe elliptica* and *Hebe salicifolia*.

Our floristic links with the Old World tropics of the Malaysian/Pacific region are confined to genera only (e.g. *Freycinetia*, *Meryta*, *Rhopalostylis* and *Earina*). There are also 35 cosmopolitan genera with a more or less world-wide distribution, for example, *Solanum*, *Gentiana*, *Ranunculus*, *Carex*, *Poa*, *Juncus*. Nearly all their New Zealand members are endemic and the few cosmopolitan species we have probably arrived here with human assistance.

During most of New Zealand's history after it broke adrift from Gondwanaland, the climate was subtropical and the land existed as a low-lying archipelago. Geologically recent events have wrought enormous changes in the appearance and biota of these islands. About five million years ago, during the Pliocene epoch, the main mountain-building phase began, and our large mountain ranges have existed in their present form for only about two million years. Superimposed on these events were the Pleistocene ice ages, which have brought about 20 warm/cold cycles during the past three million years. The last ice advance ended as recently as 10,000 years ago, and the present world climate may represent just another warm interglacial period.

The variety of new habitats created by the Pliocene mountain building apparently suited many plants — both long-established species and new arrivals from across the Tasman — as can be seen from genera such as *Hebe*, *Celmisia*, *Ranunculus*, *Raoulia*, *Aciphylla*, *Epilobium*, *Gentiana* and *Myosotis*, all of which are represented by many species, mostly in alpine regions. Debate continues as to whether existing cool-climate plants of Gondwanaland stock or trans-Tasman migrants from the mountains of Asia and Australia have been the main contributors to the 500 or so flowering plant species growing in our mountains today.

Many species became extinct during the Pleistocene glaciations, and some of the disjunct distributions seen today very likely arose when species perished in those parts of their ranges that were heavily glaciated.

One reason for the poor representation in New Zealand of large Australian families such as Proteaceae, Rutaceae, Fabaceae and Myrtaceae may be the two countries' climatic history. Before it broke adrift from Gondwanaland about 40 million years ago, Australia was linked to west Antarctica and lay under the rain shadow of India and Africa. New Zealand, by contrast, had become completely oceanic some 40 million years earlier and, when it was part of Gondwanaland, was joined to east Antarctica and fronted the Pacific Ocean. New Zealand, therefore, has probably always had a cooler, wetter climate than Australia.

The large differences between the floras of New Zealand and Australia may have more than just this historical basis, however. Dr David Lloyd (1985) suggested that there has been selection against migrant Australian species whose specialised adaptations for pollination have precluded their successful establishment in New Zealand, where there are no specialised pollinators.

The lack of specialised pollinating insects in New Zealand is also widely believed to account for the prevalence of small, white or pale flowers among New Zealand plants. However, as Dr Eric Godley (1979) pointed out, flowers which at first glance appear white may not be completely so. For example, the tubular flowers of *Euphrasia* and *Ourisia* are white outside but yellow

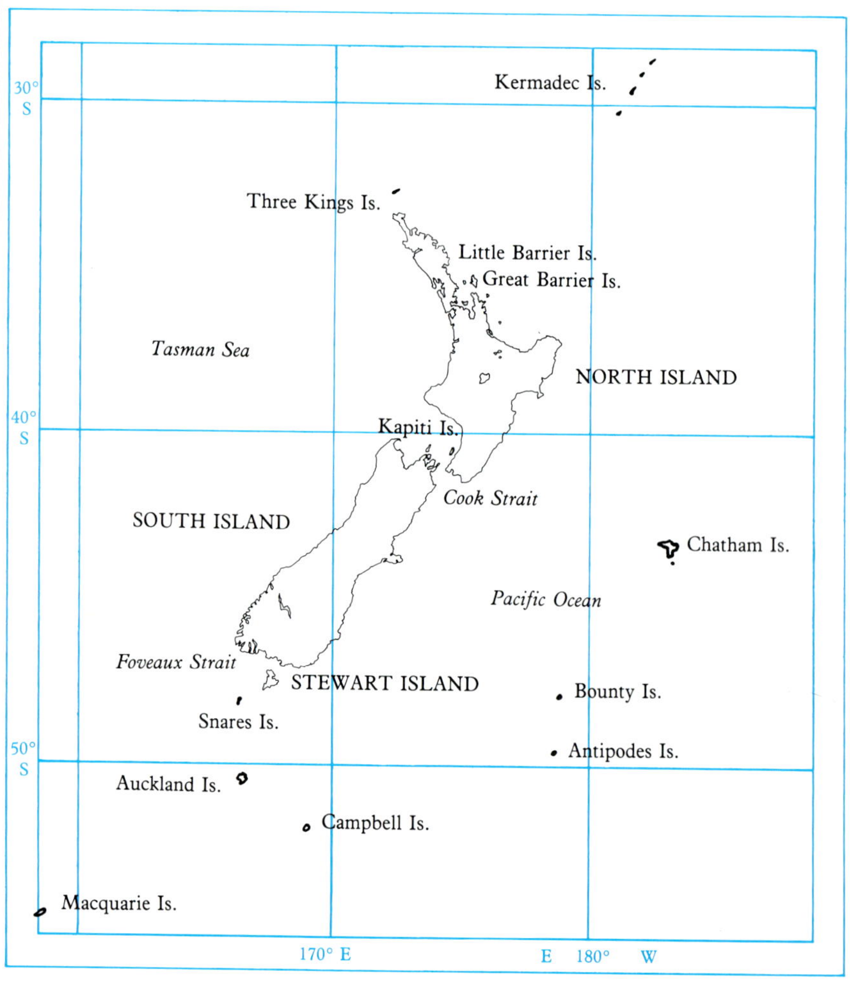

inside; hebes and parahebes often have mauve- or blue-tinted flowers; disc florets of composites are usually yellow; small purple spots, which probably serve as guide marks for insect visitors, pattern the white corollas of ngaio and *Jovellana* flowers. Lines on petals of some gentians, orchids and harebells may also function as guide marks. Guide marks are patterns of lines or spots on a petal surface that are thought to guide insects to a nectar source or, in nectarless flowers, simply help insects position themselves before landing. Not all guide marks are visible to us: research in Europe has shown that many insect-pollinated flowers have guide marks that reflect ultra-violet light and hence are invisible to humans. The occurrence on flowers of both visible and ultra-violet petal patterns was also found to be quite common. Similar extensive research has not been carried out on guide marks of New Zealand flowering plants, but there would seem to be no reason why ultra-violet petal patterns should not occur here.

Compared with the vision of humans, that of bees and many other insects is shifted along the light spectrum toward the shorter wavelengths. Thus insects can often see ultra-violet but not red light, so that a flower which appears as certain colours to us, may be perceived as quite different colours (and hence shapes and patterns) by an insect. A white flower, for example, which reflects blue, green and red light, would appear to a bee (which cannot see the red component) as a mixture of blue and green. Unspecialised flies, which do not usually have long mouthparts and so can gather nectar only from flat or concave flowers, show a preference for white, pink, yellow or green flowers. They are also attracted to reflective objects such as the nectar-filled glistening orange centres of *Corokia* flowers, or the shimmering hairs on petals or flower stalks. Beetles have poor eyesight but a good sense of smell, and the flowers they pollinate tend to be white or dull-coloured yet fragrant.

The modesty of many small flowers is frequently offset by their being massed into larger heads, for example, in *Olearia*, *Brachyglottis*, *Cassinia* and many other daisies, *Hebe*, *Parahebe*, *Pimelea*, *Weinmannia* and some species of *Dracophyllum*, *Gaultheria* and *Pittosporum*.

OFFSHORE AND OUTLYING ISLANDS

OFFSHORE AND OUTLYING ISLANDS

New Zealand is a country of islands, more than 660 of them, with all but 30 within 50 kilometres of the three largest — North, South and Stewart, which here for convenience are referred to as the mainland. Those relatively close to the mainland are offshore islands, of which the Three Kings group, 50 kilometres north of North Cape, is the largest. Further south, along the north-east coast of the North Island, between the Bay of Islands and Bay of Plenty, is a series of mainly volcanic islands such as the Poor Knights, Hen and Chicken, Little and Great Barrier, Mercury and Alderman Islands. Bare lava fields and scoria in parts of some, such as Mayor and Rangitoto, indicate their recent volcanic origin, while one, White Island, is still active, its vapour and ash clearly visible from the mainland.

Other offshore islands are situated around Cook Strait, the Marlborough Sounds, south-west Fiordland, Foveaux Strait and Stewart Island.

Outlying islands are those more than 50 kilometres from the mainland. They range from the subtropical and actively volcanic Kermadec group, nearly 1000 kilometres north of Auckland, through the cloudy but equable Chathams, 800 kilometres east of Christchurch, to the gale-lashed subantarctic islands that dot the Southern Ocean between 250 kilometres (the Snares) and almost 1000 kilometres (Macquarie Island) from the southern South Island.

Apart from the large average temperature differences between the sub-

Pleurophyllum speciosum and *Stilbocarpa polaris* at 'Fairchild's Garden', Adams Island, Auckland Islands.

R.M.

Stilbocarpa polaris, Adams Island, Auckland Islands group.

R.M.

A. & S.P.

South Harbour, Poor Knights Islands.

tropical and subantarctic outlying islands, the plants of these and all those between are influenced primarily by the proximity of the sea itself, and by salt-laden winds. The powerful effects of wind may be seen on the Chatham and subantarctic islands where forest growth is dwarfed to that of dense, wind-trimmed scrub.

The narrow genetic base from which island plants have evolved has given rise to new races and species that are closely related to, yet quite distinct from, their mainland cousins. We have, for instance, the Poor Knights ngaio (*Myoporum laetum* var. *decumbens*), the Kermadec tree fern (*Alsophila kermadecensis*), five-finger (*Pseudopanax arboreus* var. *kermadecensis*) and pohutukawa (*Metrosideros kermadecensis*), the Three Kings rangiora (*Brachyglottis arborescens*), cabbage tree (*Cordyline kaspar*) and titoki (*Alectryon grandis*), nearly 40 endemic races and species on the Chatham Islands, most of which are closely related to mainland taxa.

As well as endemic varieties and species closely related to mainlanders, our offshore and outlying islands also have a disproportionately large share (in relation to their area) of endemic genera and species with no close mainland relatives. One important reason for this is believed to be that islands served as refugia for species which became extinct on the mainland, because of, for example, changes in ice advance, volcanic eruptions, fire or disease. Included are representatives from the ancient Gondwanaland floras, such as *Myosotidium* and *Pleurophyllum*, as well as those present in floras of the subtropical early to mid-Tertiary period, such as *Meryta*, *Xeronema* and *Tecomanthe*.

Most of New Zealand's larger islands have been considerably modified by human activity, while a very few remain in a near-pristine condition. On the Chathams, with their long history of settlement and pastoral farming, habitats of native plants have been progressively destroyed, to the extent that now nearly half the islands' 37 endemic taxa are threatened with extinction. More fortunate have been a few islands in the Kermadec group, in south-west Fiordland, and among the subantarctic groups, where there is

P.M. & J.L.S.

Hebe bollonsii, Poor Knights.

no human settlement, and where neither domestic animals nor rodents have been introduced. Others, such as the Three Kings and Campbell Islands, are recovering from earlier severe depletion now that introduced grazing animals have been removed.

Pollinators of island plants include the same wide range of birds and unspecialised insects that are characteristic of the mainland flora. However, among the more unusual visitors to northern offshore island plants such as the Poor Knight's lily (*Xeronema callistemon*) and pohutukawa (*Metrosideros excelsa*), are the nocturnal geckos (*Hoplodactylus* spp.). Some biologists believe that before mainland lizard populations were decimated by rats, lizards were important pollinators of species such as these, whose flowers have upright stamens and exposed pollen. As both geckos and skinks have been observed eating some fleshy fruits e.g. kawakawa (*Macropiper excelsum*) and taupata (*Coprosma repens*), it is possible that they were also important distributors of certain species.

Reasons for the more colourful flowers of island plants, compared with those of their mainland relatives, remain a matter for conjecture, although several theories have been proposed. In 1904 Leonard Cockayne suggested that the brighter flowers of southern outlying island plants arose in response to longer exposure to sun during long summer days. However, this theory does not account for the variable flower colour in several southern island plants, such as gentians, and the relatively low mean annual sunshine hours these islands receive.

Present-day botanists have other theories: Dr Peter Wardle (1978) suggested that showy-flowered island plants were relicts of a cool-climate flora that existed in southern New Zealand during the Tertiary and which also included the ancestors of many present-day alpine species.

Dr Eric Godley (1979) took the view that coloured flowers in island plants represent selection for any features increasing a plant's chances of pollination in environments where pollinators are scarce. Perhaps conspicuous flowers also serve as beacons to pollinators on windy subantarctic islands — helping them to make more direct, energy-conserving trips between the flowers they visit. However, even if the reasons for their prevalence remain unresolved, the colourful flowers of many island plants stamp our flora with a unique and spectacular beauty.

R.M.

Brachyglottis huntii, Pitt Island, Chathams group.

R.M.

Aciphylla dieffenbachii

Aciphylla dieffenbachii
Dieffenbach's speargrass

The umbrella-shaped flower heads of this rare Chatham Island herb suggest its alliance to the mainland speargrasses and other plants of the carrot family. Male and female flowers occur on separate plants, and although individually tiny, the massed cream flowers, in clustered umbels at the top of each flower stem, attractively complement the bluish-green hairless leaves.

Up to a metre tall, Dieffenbach's speargrass contrasts with many of its mainland cousins, such as *Aciphylla aurea*, *A. colensoi* and *A. scott-thomsonii*, in having flaccid, soft-tipped leaves.

Leaves are up to 600mm long and arise directly from the plant base. They are divided into two or three series of narrow, linear segments, which range from 8mm down to 2mm wide.

Like many other Chatham Island endemics, wild populations of *Aciphylla dieffenbachii* have been decimated by browsing stock, so that it is now confined to inaccessible coastal cliffs.

Its dioecious breeding system, whereby male and female flowers are on separate plants, further hampers replenishment of wild populations: only half the plants (i.e. females) will produce seed, and isolated plants are unlikely to do so. However, once formed, seed germinates well and this, together with the species' ornamental appeal, should ensure its survival in horticulture, at least.

Chatham and Stewart Islands — coastal cliffs and rocky slopes on the Chatham Islands and also naturalised at a few sites in Stewart Island.

UMBELLIFERAE

R.M.

C.E.B.

Anisotome latifolia

Anisotome antipoda
Anisotome latifolia

These large subantarctic herbs are a spectacular feature of coastal cliffs and herbfields on Auckland and Campbell Islands. Unlike their cream-flowered mainland cousins, they have tiny pale to deep pink flowers arranged in large cauliflower-like compound umbels borne on each stout, grooved flower stalk, which may be up to one metre tall.

Both species have the pinnately compound leaves typical of carrot family herbs, but those of *A. antipoda* are very finely divided and are the smaller of the two (up to 400mm long compared with 600mm in *A. latifolia*). Leaves of both species are leathery textured and have especially thick waxy cuticles, which help the plants withstand the damaging and desiccating effects of salt-laden gales.

These once common herbs have suffered greatly through their palatability to cattle, sheep, pigs and rabbits, which were introduced to these islands, and *A. latifolia* now is mostly confined to cliffs and other places inaccessible to browsing animals.

Auckland and Campbell Islands — coastal cliffs and herbfield.

UMBELLIFERAE

A. & S.P.

Brachyglottis huntii

R.M.

Brachyglottis stewartiae

R.M.

Brachyglottis huntii

Brachyglottis huntii
Rautini

With panicles of rich yellow flower heads crowning its dense, rounded canopies (to eight metres tall), rautini's robust good looks delight Chatham Islanders each summer. There it is dubbed the 'Chatham Island Christmas tree', because of its predictable flowering in late December and early January.

The foliage too has a singular charm: large (50–100 by 20–30 mm), mid-green leaves clustered at the branchlet tips are tacky and hairy while young, but become almost hairless as they age. Other distinctive features are the leaves' fresh lemon scent and rolled-down margins.

Formerly rautini was much more numerous and widespread on the Chatham Islands, but grazing and pasture development have now confined it to relatively small areas on Pitt and Chatham Islands. Long-term, its survival in the wild will depend on the establishment of fenced reserves where seedlings can grow without being eaten or trampled by stock.
Chatham Islands — forest and scrub on gully edges, hillsides and cliffs on Pitt Island and the southern tablelands of Chatham Island.

COMPOSITAE

Brachyglottis stewartiae

A close relative of *Brachyglottis huntii*, *B. stewartiae* is confined to offshore islands of southern New Zealand — Stewart Island, the Snares and Solander Islands and islets in Foveaux Strait.

Large elliptic leaves (up to 180 mm long and 40 mm wide) are clustered at the ends of pale, stout branches, which from December to about February bear showy, dense panicles of yellow flower heads.

The growth habit in the wild is rather curious: initially upright, the shrub's repeated branching, combined with frequent gales and petrel burrows (which destroy the cohesiveness of the peaty soil), causes it to topple over. Once prostrate, the stems develop adventitious roots to anchor the plant in the soil. The densely branching habit produces a very tight leafy canopy.
Stewart, Snares, Solander Islands and islets in Foveaux Strait — coastal forest and scrub.

COMPOSITAE

R.M.

Bulbinella rossii

R.M.

Carmichaelia williamsii

D.G.

Bulbinella rossii

Bulbinella rossii

Small, bright golden-yellow flowers are densely crowded in racemes up to 150mm long and 60mm wide, borne on the ends of stout, 300mm long flower stalks. Male and female flowers occur on separate plants, the male inflorescences larger than the female.

Like several other species found only on our subantarctic islands, *Bulbinella rossii* is larger and showier than its mainland relatives. Its very attractive flowers, which usually appear around December and January, are teamed with bright green, fleshy, strap-shaped leaves (up to 300 by 60mm), which are borne on stems up to one metre tall. Outer leaves curve outwards, while the inner stand more erect, and all are strongly keeled, being V-shaped in cross-section. The leaves die down in winter, the plant perennating from a fleshy rootstock rich in food reserves.

Its summer-green habit and unpalatability to cattle, rabbits and pigs have meant that *Bulbinella rossii* has not been eaten out to the same extent as many other beautiful herbs of our outlying islands.

Auckland and Campbell Islands — herbfield.

LILIACEAE

Carmichaelia williamsii

Our most easily recognised native broom, *Carmichaelia williamsii* has much wider flat photosynthetic branchlets (to 15mm) and larger flowers (to 25mm long) than other native brooms. Flower colour, too, is atypical, the petals being mainly yellow, rather than mauve, although they do have purple veins.

When dry, the two sides of the chubby 25mm long pods fall off to reveal half a dozen or so bright orange seeds suspended from the pod rim.

Now dwindling in numbers in the wild, there is probably only one population of *Carmichaelia williamsii* remaining on the mainland.

Despite its northerly distribution, the species thrives in cultivation far south of its natural range, at least if given a sheltered, frost-free site. Its popularity as an ornamental and its ease of propagation from seed and cuttings will probably ensure its survival even if wild populations disappear. In Dunedin, plants flower during winter and early spring and the pods open by autumn.

North Island and offshore islands — coastal areas from the Bay of Plenty to East Cape and on the Poor Knights, Little Barrier and Alderman Islands, but now rare in the wild.

LEGUMINOSAE

C.P.

Corokia macrocarpa

Corokia macrocarpa

Whakataka

With the advent of spring, whakataka is transformed by the opening of its racemes of small (10 mm diameter), starry, yellow-petalled flowers, each having a central orange disc gleaming with nectar. This attractive and prolific display continues over four weeks or more, hence the popularity of this species as an ornamental.

Its neat form and two-tone foliage (mid-green above, silvery felted below), tolerance of clipping and of a wide range of soil types and situations, further recommend whakataka to the gardener.

By autumn the flowers have matured into barrel-shaped, yellowish-orange drupes each 10 mm wide and containing a single stony seed.

The largest of New Zealand's three corokias, whakataka sometimes reaches small tree size at six metres or more.

Although it is not as rare as many other Chatham Island endemics, the fortunes of *Corokia macrocarpa* have suffered badly as a result of forest clearance and pasture development. However, its prospects for survival look brighter than those of many other Chatham Island plants because of its horticultural merit.

Chatham Islands — forest and scrub.

CORNACEAE

Geranium traversii R.M.

Geranium traversii

Chatham Islands geranium

A long-standing favourite of New Zealand rock gardeners, the large (25 mm across), pink flowers and silky-hairy, grey-green leaves of this rosette herb make it an irresistible acquisition for the enthusiast.

It perennates by means of a stout taproot, which often gives rise to arched runners that send up offshoots. The long (up to 200 mm), slender petioles produce round but deeply lobed and toothed leaves, which are covered in silky, silvery hairs above and below.

The form of *Geranium traversii* most often seen in cultivation has deep pink flowers whereas wild populations on the Chatham Islands contain both white- and pink-flowered plants. *Geranium traversii* certainly monopolises the horticultural merits of its genus in New Zealand — none of the other four species of native geraniums approach it in showiness.

Although certainly less common than it once was in the wild, *Geranium traversii* has fared rather better than many other Chatham Island endemics, partly because it propagates vigorously.

Chatham Islands — coastal cliffs and sand dunes.

GERANIACEAE

D.G.

Hebe benthamii

Hebe benthamii

This stout, loosely branched shrub grows from 200 to 500 mm tall in rocky places amongst tussocky grasses, shrubs and herbs on the Auckland and Campbell Islands. It vies with such notable cousins as *Hebe speciosa* and *Hebe macrantha* as the most attractive member of its genus; its bright blue petals are unique among New Zealand hebes and are borne in 100 mm long racemes crowded at the branch ends. Each flower is up to 8 mm across and is unusual in having five or six corolla and calyx lobes rather than four. Fruit capsules are also atypical in having three rather than the customary two compartments.

Margins of the leathery leaves (which are up to 25 by 10 mm) are notched and fringed with hairs. Plants have a somewhat leggy appearance: leaves are confined to the smaller branchlets, while the older, larger branches are bare and prominently marked with circular leaf scars.

Auckland and Campbell Islands — rocky sites in damp shrubland.

SCROPHULARIACEAE

J.F.

Hebe chathamica

Hebe chathamica

Chatham Islands koromiko

Found only on Chatham Island coasts, this metre-wide trailing shrub is highly variable in growth habit, leaf and stem morphology and flower colour. Its variability probably arises from its ability to hybridise with another Chatham Island endemic, *Hebe dieffenbachii.* Branchlets may be red or green and, like the leaves, hairy or glabrous. The thick, stalkless leaves tend to lie in one plane and are generally broadly oblong, ranging from 15 to 36 mm long by 5 to 12 mm wide. Flower colour varies from white to lilac, the small flowers being borne in compact spikes about 40 mm long and almost as broad.

Chatham Islands — exposed rocky coastal sites, sometimes forming dense mats in saltmarsh as well.

SCROPHULARIACEAE

D.G.

Hebe chathamica

R.M.

Myosotidium hortensia

Myosotidium hortensia

Chatham Islands forget-me-not

Once common around the coasts of most islands in the Chatham group, this regal species is now confined to just a few sites. Its decline has been caused mainly by stock eating and trampling the plants and it now survives only where it is inaccessible to animals — on cliff ledges and other such coastal sites.

Although related to the common forget-me-nots, *Myosotidium* is distinct enough to warrant placement in a separate genus, of which it is the only member. It is a spectacular perennial herb, sometimes spanning more than a metre. The huge broad leaves, each 300 mm or more wide, are prominently veined and shiny, and are borne on stout fleshy stalks up to 500 mm long.

Flowers are a fitting match for the foliage: clustered in dense pyramidal heads up to 150 mm across, they feature numerous 15 mm wide sky-blue florets, shading to deep purplish-blue at their bases. Occasionally white-flowered plants occur, and are a sought-after addition to any native garden. Flowers develop into the characteristic 15 mm wide, flat fruits, each surrounded by a wide wing.

Chatham Islands — coastal sand dunes and beaches.

BORAGINACEAE

R.M.

Myosotidium hortensia

M.F.S.

Myosotis capitata

Myosotis capitata

Auckland Islands forget-me-not

This little-known species, which occurs only on the subantarctic Auckland and Campbell Islands, is perhaps our showiest native *Myosotis*. Single rosettes of mid-green, strap-shaped leaves (up to 120mm long and 25mm wide) are much less hairy than those of other native forget-me-nots. From November, lovely deep blue cymes of flowers are produced, each cluster up to 50mm wide. Each flower's tubular corolla is about 10mm long, flaring into five lobes at the mouth.
Auckland and Campbell Islands — rocky sites.

A.&S.P.

Olearia chathamica

Olearia chathamica

Keketerehe

This beautiful Chatham Island shrub daisy is a close relative of the South and Stewart Island species *Olearia angustifolia* and *O. oporina*. Stock grazing has resulted in its now being rare on Chatham Island, the only island in the group where it occurs.

Olearia chathamica grows into a shrub of about two metres, or sometimes into a small tree, preferring forest margins in well-drained coastal sites, although it occasionally occurs in forest.

The very leathery, serrate leaves are a glossy dark green above and are coated beneath with a felt of white hairs. The showy flower heads (each up to 45mm diameter) are borne in groups and are very conspicuous with their white to purple outer ray florets and purple inner disc florets.
Chatham Island — coastal cliffs, forest, forest margins and scrub.

J.F.

Olearia chathamica

R.M.

Olearia semidentata

Olearia semidentata

Chatham Islands aster

This eye-catching metre-tall shrub is cast in a finer mould than its Chatham Islands compatriot *Olearia chathamica*, having more slender branches and narrower leaves, which are toothed in their upper part, dark green and glossy above and white tomentose below. In flower it is most attractive, with 35 mm wide, pink to violet-rayed heads with purple centres.

Like *O. chathamica*, it is now rare in the Chatham Islands because its habitat — wet, peaty shrubland — continues to be developed for grazing. It often grows in association with the giant rush, *Sporodanthus traversii*, and where protected from trampling stock, seedlings rapidly colonise burnt areas.

In keeping with their aristocratic good looks, this species and *O. chathamica*, as well as the South and Stewart Island species, *O. angustifolia* and *O. oporina*, are all difficult to grow in cultivation, being prone to sudden collapse, especially in summer.

Chatham Islands — shrubland on wet, peaty ground.

COMPOSITAE

M.F.S.

Pleurophyllum criniferum

Pleurophyllum criniferum

This summer-green herb has longer, thinner, less markedly ribbed and more erect leaves, which vary more in shape than those of *Pleurophyllum speciosum*. (Leaves may be up to one metre long and from 150 to 300 mm broad.) Leaf undersurfaces are covered in white cottony hairs and are strengthened by about 20 stout flexible ribs. The leafy, bracted flower stalks are massive (up to two metres tall) and bear racemes of 15 to 30 heads (40 mm wide), which comprise numerous purplish-brown tubular disc florets.

Auckland, Campbell and Antipodes Islands — coastal and lower montane peaty herbfield.

COMPOSITAE

M.F.S.

Pleurophyllum speciosum

R.M.

Stilbocarpa polaris

D.G.

Pleurophyllum speciosum

Pleurophyllum speciosum

One of the most stately species in the flora, *Pleurophyllum speciosum* excels in both flower and foliage. Although it is allied to *Olearia*, the leaves are reminiscent of those of the larger hostas — obovate to rhomboid in shape, thick and bright green, with numerous prominent parallel ribs. They may be up to 600mm long and 400mm wide, with the four to five leaves of each plant arranged in a cup-like rosette. Leaf ribs are most prominent on the undersurface and the furrows between them are filled with cobwebby pale hairs.

The flowers, however, reveal this herb's daisy family connections: about 25 white-rayed, purple-centred heads (each 50 mm across) crown stout, hairy stalks up to a metre tall.
Auckland and Campbell Islands — coastal to montane herbfield.

COMPOSITAE

Stilbocarpa polaris

Several subantarctic islands are the home of this stately herb, which may be up to two metres across. Much-branched stems, arising from a thick rhizome, bear coarsely hairy, rather corrugated-looking, kidney-shaped leaves, which are wider (450 mm) than they are long (300mm) and are borne on stalks up to 600 mm long.

About November, masses of tiny, waxy, yellow flowers are produced in globose compound umbels, which are up to 300mm across and sit atop a stout hairy stalk about 150mm long. These are succeeded by shiny, black, round berries, each about 5mm in diameter.

Stilbocarpa polaris was dubbed the 'Macquarie Island cabbage' by early sealers and whalers, who ate the leaves and leaf stalks to prevent scurvy. Palatable not only to humans but also to the animals introduced to those islands, this plant has been badly eaten out in some areas.
Auckland, Campbell, Antipodes and Macquarie Islands — coasts, herbfield and shrubland.

ARALIACEAE

P.M. & J.L.S.

Stilbocarpa robusta

Stilbocarpa robusta

This Snares Islands endemic has bright green, kidney-shaped leaves up to 600 mm across, which are less hairy and more deeply lobed than in *Stilbocarpa polaris*. Stout stalks up to 500 mm long send up round bracted heads comprising many small pale yellow flowers.

Snares Islands — open forest, clearings and scrub.

ARALIACEAE

R.M.

Tecomanthe speciosa

Tecomanthe speciosa

A single plant on Great Island in the Three Kings group is all that survives of this species in the wild, the others presumably having been eaten by goats, which were introduced to the Three Kings last century.

Professor G. T. S. Baylis discovered this lone wild plant on Great Island in 1945, and in 1951 brought cuttings back to the Mt Albert Research Centre of the DSIR in Auckland. Seed was obtained after those cuttings flowered and all cultivated plants thus derive from the specimen on Great Island.

Fortunately *Tecomanthe* is a vigorous climber and is very amenable to cultivation, growing up to several metres a year and propagating readily from seed and cuttings. It may flower in as little as three years when grown from a cutting.

A member of the tropical bignonia family, its closest relatives are some 20 other species of *Tecomanthe*, which occur mainly in New Guinea. The handsome glossy foliage is rather like that of kohekohe (*Dysoxylum spectabile*). The leaves are compound, comprising three to five pairs of lateral leaflets and a terminal leaflet, all more or less broadly elliptic in shape, and with the terminal leaflet slightly larger (to 120 by 90 mm) than the others.

Dense clusters of about 30 creamy,

R.M.

Tecomanthe speciosa

green-flushed flowers appear from May through August. Hairs on the outside of the petals give the flowers an unusual velvety texture. Each tubular flower is about 60 mm long and opens to a five-lobed mouth. Flowers are succeeded by woody pods up to 180 mm long, which split from the base into two canoe-shaped valves containing rows of 200 or more seeds, each with a thin, broad wing.

Great Island in the Three Kings group. Plants are now available through the nursery trade.

BIGNONIACEAE

R.M.

Xeronema callistemon

R.M.

Xeronema callistemon

Xeronema callistemon

Raupotaranga, Poor Knights lily

This rare and beautiful lily family plant was discovered on the Poor Knights Islands around 1920 by W. M. Fraser of the Whangarei Harbour Board. It was classified by Dr W. R. B. Oliver of the then Dominion Museum, who found that the plant's closest relative was *Xeronema moorei*, which grows on the mountains of New Caledonia. Despite their large geographical separation, this and New Zealand's *Xeronema* are remarkably similar.

Xeronema translates as 'dry threads', in reference to the tepals and stamens of the spent flower spike. The New Zealand *Xeronema* was given the specific name *callistemon*, meaning 'beautiful stamen' — apt, because the showiness of the long inflorescence comes from the conspicuous red stamens. *Callistemon* is also the generic name of Australia's bottlebrushes, shrubs botanically quite unrelated to *Xeronema* and belonging in fact to the eucalypt and rata family but also having flowers with showy stamens.

When not in flower, *Xeronema callistemon* looks very like a small flax with its fan-like, sword-shaped clumps of rigid leaves, each up to one metre long and 50 mm wide. Around October a stout flower stalk appears from the centre of the leafy fan. This eventually arches over the leaves and bears a dense spike of red flowers up to 350 mm long, conspicuously arranged on its upper side. (Flowers are actually produced all around the main flower stalk, but the stalks of individual flowers on the underside bend upwards.)

Xeronema flowers lack petals, their place being taken by narrow, red tepals which hang below the pistil, while the six stamens surrounding the pistil point upwards. By as early as December in some plants, the flowers have matured into dry, brown capsular fruits, whose valves gape apart so that the black spiny seeds (similar to those of rengarenga) fall out.

In cultivation, *Xeronema* spurns the lavish attention often heaped on it, thriving best in a well-drained, soil-less mix in a porous, tight-fitting container. These preferences are in keeping with those in the wild, where it thrives in rocky, windswept ledges with a minimum of soil.

Poor Knights and Hen (Taranga) Islands — coastal rocks and cliffs.

COASTAL OPEN PLACES

Metrosideros excelsa, Poor Knights coast.

Metrosideros excelsa flowers.

R.M.

COASTAL OPEN PLACES

New Zealand's 10,000-kilometre coastline embraces a diversity of topography and habitats. Landforms are as varied as the vast dunes of Ninety Mile Beach, the spectacular sheer cliffs of Banks Peninsula and the drowned, glacier-hewn valleys of Fiordland. Within these, open habitats may range from cliffs and rocky foreshores to sand dunes, saltmarsh and shrubland.

Many of us live near the coast and are familiar with our local seashores as summer havens for recreation. Most likely, though, we know the coast only as fine-weather friends — unaware of the severity of wind and wave it can experience at times. Coastal plants must be able to withstand and thrive through all the vagaries of weather, as well as surmount the problems peculiar to their situation. These latter include salt-laden winds, high concentrations of salt in the soil, intense exposure to both direct sunlight and the reflected light from ocean and estuary, burial by mobile sand and, for some, twice-daily inundation by the tide. Yet the proximity of the sea does buffer against extreme cold, and severe frosts are rare, except on the coast near the Canterbury Plains, where temperatures may drop as low as −9°C.

R.M.

R.M.

Hebe elliptica.

Salt-laden wind is the most dominating influence for all coastal plants, whatever their niche. The wind-trimmed, 'flag form' shapes of many seashore trees and shrubs testify to this. A number of species, for example, horokaka (*Disphyma australe*), *Selliera radicans*, and bachelor's buttons (*Cotula coronopifolia*), have developed succulent leaves or stems to overcome excessive salt in air and soil. Others have a thicker than normal waxy coating or cuticle to protect their leaves, for example, taupata (*Coprosma repens*) and ngaio (*Myoporum laetum*). Some, such as pingao (*Desmoschoenus spiralis*), minimise water loss by having thick, rigid leaves. A number — for instance, creeping fuchsia (*Fuchsia procumbens*), nihinihi (*Calystegia soldanella*) and *Muehlenbeckia complexa* — simply avoid the wind by growing as low mats or trailers.

Among all these diverse solutions to the challenges of a coastal environment are the flowers — some conspicuous and colourful, others more modest but pervasively fragrant. The following selection includes both common species, found throughout our shores, and others that are rare, with very restricted ranges.

The following also occur in coastal open places: *Clianthus puniceus, Corokia cotoneaster, Hymenanthera alpina, Lagenifera pumila, Olearia avicenniifolia, Phormium cookianum, Phormium tenax, Rhabdothamnus solandri.*

P.M. & J.L.S.

Arthropodium cirratum

R.M.

Arthropodium cirratum

Arthropodium cirratum

Rengarenga, rock lily

Some rengarengas can usually be found in flower any time between July and December. The mauve-tinted buds open to white petals with just a hint of mauve on their undersides. The specific name, *cirratum*, comes from the Latin *cirratus*, meaning 'curled', and refers to the yellow, upward-curved tails of the purple, fuzzy anthers, which under a hand lens look rather like miniature bottlebrushes.

Flowers are borne in decorative, lax, metre-tall panicles, and by late summer have ripened into capsules containing numerous hard seeds, each about 2mm across and with minutely embossed black coats.

This handsome herb, with its mid-green, partly drooping, rather flax-like leaves (up to 600mm long and 65mm wide), grows wild on sea cliffs from North Cape to the Marlborough coast of Cook Strait. It can, however, grow and seed happily in the southern South Island; despite the frost tenderness of its foliage it can usually survive cold spells because of its stout underground rhizome.

North and South Islands — coastal cliffs and rocky places from North Cape to the Marlborough coast of Cook Strait.

LILIACEAE

G.R.R.

Calystegia soldanella

Calystegia soldanella

Shore convolvulus

A creeping, branching rhizome bears the glossy, rather fleshy, kidney-shaped leaves of this sand-dune plant in patches up to two metres or more across.

Flowering begins in October in warmer localities, or as late as March in colder sites. The white to purplish-pink flowers are relatively large (up to 50mm across) and conspicuous, sitting atop the mat of foliage on stalks up to 100mm long. The five petals are joined along their edges into a five-angled, funnel-shaped corolla, which has five pale primrose bands down its throat.

Kermadec, North, South, Stewart and Chatham Islands — coastal sands or occasionally gravels and sandy inland lake shores.

CONVOLVULACEAE

P.N.J.

Cotula coronopifolia

A. & S.P.

Disphyma australe

Cotula coronopifolia

Bachelor's buttons, yellow buttons

Familiar to almost everyone who has visited the seashore in summer, mats of this rather succulent perennial herb then sport numerous bright yellow, rayless, button-like flower heads, each about 10mm across. It occurs in many temperate parts of the world and is one of the relatively few cosmopolitans in our flora.

The rather fleshy, hairless leaves vary in shape — some are sharply lobed, others smooth edged. They are usually between 50 and 80mm long and their bases encircle the stem.

Three Kings, North, South, Stewart and Chatham Islands — coastal to lowland muddy shores, damp sand hollows, bogs and swamp margins. Also Europe, South Africa, Australia and South America.

COMPOSITAE

Disphyma australe

Horokaka, native ice plant

A familiar sight on coasts throughout New Zealand, horokaka's trailing stems can often be seen festooning several metres down cliffs. Its paired, turgid, three-cornered leaves contain special water-storage tissue. Leaf colour varies from green to red, probably owing to both genetic and environmental differences. Although red- and green-leaved forms often occur together, in sunny, exposed sites leaves are more often bright pinkish-red.

The showiness of the 30mm wide flowers comes from numerous white to pink staminodes (or non-functional stamens). Horokaka flowers have no petals. Flowers may appear any time from July to March and mature into initially succulent, but later dry, hard capsules. These open only in high humidities, so that seed release occurs when there is ample moisture for germination.

Records in the collected journals (1833–73) of the Reverend Richard Taylor show that the Maori used juice squeezed from horokaka leaves to treat boils.

Kermadec, North, South, Chatham and Stewart Islands — coastal rocks and cliffs.

AIZOACEAE

J.B.

Disphyma australe

R.M.

Euphorbia glauca

Euphorbia glauca

Shore spurge

Although threatened with extinction in the wild, its ease of cultivation will probably save shore spurge from total extinction, even if wild populations continue to diminish.

Blue-green leaves and reddish half-metre-tall stems contrast strikingly with maroon-edged bracts surrounding the several apetalous male and single female flowers enclosed within a floral cup. Three-valved capsular fruits hang over the edge of the floral cup, and when ripe the capsules split, flinging out several pale seeds.

Although shore spurge is our only native *Euphorbia*, world-wide the genus is a huge one, with about 2000 species including familiar annual weeds such as milkweed (*Euphorbia peplus*) and perennial ornamental herbs such as poinsettia (*Euphorbia pulcherrima*). The para rubber plant (*Hevea brasiliensis*), which is the source of most of the world's rubber, is also a member of the euphorbia family.

Norfolk, North, South and Stewart Islands — coastal sands and rocks.

EUPHORBIACEAE

Fuchsia procumbens C.E.B.

Fuchsia procumbens

Creeping fuchsia

In contrast to its cousin the tree fuchsia, which is the largest species in its genus, *Fuchsia procumbens*, the world's smallest fuchsia, is a herb with weak scrambling stems. Such a habit makes it an attractive subject for cultivation in hanging baskets and as a ground cover. Its flowers, as well as having blue pollen like *F. excorticata*, are unusual in that they lack petals, stand erect (unlike the pendent flowers of other fuchsias) and develop into surprisingly large (20 mm long), scarlet berries.

In danger of becoming extinct in the wild, *F. procumbens* is known from only a few localities between the Coromandel Peninsula and North Cape. Even these few remaining populations are threatened by erosion of coastal sites and competing exotics such as kikuyu and buffalo grasses.

North Island — coastal and lowland sandy and rocky places, scrub and grassland from North Cape to Coromandel Peninsula.

ONAGRACEAE

R.M.

Hebe elliptica

Hebe elliptica

Even the buffeting salt-charged winds of exposed southern and western coasts cannot deter this coastal koromiko from staging an impressive floral display each summer. The mauve to white flowers are relatively large for the genus (up to 8 mm across) and are borne in short axillary clusters of four to fourteen.

Hebe elliptica forms a densely branched and leafy shrub up to three metres tall — groups on exposed cliffs can often look like a wind-trimmed hedge. It is exclusively coastal and in many areas is the dominant shrub.

North, South, Stewart and subantarctic islands — cliffs on western coasts from Mt Egmont south, and eastern coasts from Otago south. Also the Falkland Islands and southern Chile.

SCROPHULARIACEAE

R.M.

Hebe speciosa

Hebe speciosa
Titirangi, napuka

The species name means 'beautiful' and was given by Richard Cunningham when he discovered this eye-catching shrub at Hokianga Heads in 1834.

Hebe speciosa has the added attraction of flowering through late summer, autumn or winter, and some of the garden hybrids derived from it have inherited this trait.

Short dense spikes (to 100 by 30 mm) of deep magenta flowers are teamed with thick, red-margined, glossy leaves each up to 100 by 50 mm.

Browsing by stock and coastal erosion have led to the reduction of wild populations of this species, but its showiness should ensure a continued popularity in cultivation.

North and South Islands — coastal cliffs in a few scattered sites in the western North Island, from Hokianga Heads to northern Taranaki, and also in the Marlborough Sounds.

SCROPHULARIACEAE

R.M.

Hibiscus trionum

Hibiscus trionum
Native hibiscus, puarangi

A relatively hardy species for a member of a largely tropical genus, this attractive half-metre-tall annual or biennial herb produces a succession of large (50 mm diameter) pale primrose, black-throated flowers throughout summer and autumn. These are followed by conspicuously veined, pale, papery seed capsules.

Although rare in the wild, it naturalises freely from seed in warm sites throughout the country, even in the southern South Island. There is some evidence that *Hibiscus trionum* may have been introduced to New Zealand by the Maori who used the leaves for cleaning hands and may have cultivated it for this purpose and for its attractive flowers.

North and South Islands, and offshore islands — sheltered usually coastal sites mainly around North Auckland and nearby offshore islands, but naturalised widely through much of the two main islands.

MALVACEAE

R.M.

Hibiscus trionum

D.G.

Linum monogynum

G.R.R.

Metrosideros excelsa

Linum monogynum

Rauhuia

This New Zealand relative of the European and Asian linen flax can be found in rocky places, usually on or near the coast. Although largely herbaceous, rauhuia has a soft woody base. Bushy stems bear stiffly ascending branches, and plants may be up to half a metre tall. Despite being unrelated botanically to New Zealand flax (*Phormium tenax*), rauhuia does have in common with it an abundance of fibrous tissues.

Flowers appear from October in warmer districts, and from December in cooler areas further south. The white flowers are up to 30 mm across and are borne in clusters at the branch tips, making a conspicuous show. They mature into ovoid capsules about 10 mm across, containing numerous flat, glossy brown seeds.

Three Kings, North, South, Stewart and Chatham Islands, also offshore islands — rocky sites and dunes in coastal and lowland open places.

LINACEAE

Metrosideros excelsa

Pohutukawa

Superlatives have often been used to describe pohutukawa's spectacular crimson flowers: for Thomas Kirk, writing in *The Forest Flora of New Zealand* (1899), they made pohutukawa '. . . perhaps the most magnificent plant in the New Zealand flora'. To early European settlers pohutukawa was the 'Christmas tree', whose showy flowers were conveniently timed for Christmas decorations.

Clinging to rocky cliffs by its long, twisted roots and net-like tangles of red, fibrous aerial roots, pohutukawa is truly at one with the sea, and branches that reach the inter-tidal zone have been found with oysters clinging to them.

The generic name derives from the Greek *metra*, 'middle', and *sideros*, 'iron', in reference to the hardness of the heart wood. Pohutukawa's dense, dark red wood is very strong and durable and was formerly used in boatbuilding.

Although closely related to rata (*Metrosideros robusta*), and sometimes even hybridising with it, pohutukawa is distinguished by its larger, broader leaves with their pale, hairy undersides. Pohutukawa hybridises readily with its Kermadec Island cousin (*Metrosideros kermadecensis*), which also has hairy leaf undersides but smaller leaves, and such hybrids are sometimes sold in the nursery trade.

Three Kings and North Islands — open coastal sites and coastal forest as far south as Taranaki and Poverty Bay. Those around lake shores on the Volcanic Plateau may have been planted by the Maori.

MYRTACEAE

R.M.

Yellow-flowered cultivar of *Metrosideros excelsa*

J.F.
Mimulus repens

Mimulus repens
Native musk

Once much more common than it is today, owing to habitat depletion through reclamation of saltmarshes and competition from introduced weeds, *Mimulus repens* is still abundant in a few coastal localities throughout the North and South Islands.

Mats of succulent, pale, creeping stems bear small (5 mm long) leaves. From October to February (depending on location), these mats are covered in numerous small (12 mm across) flowers, which are mauve without but have a yellow throat and the frilled, flaring petals typical of the snapdragon family.

North and South Islands — coastal saltmarshes and damp sands. Also Tasmania.

SCROPHULARIACEAE

R.M.
Myoporum laetum

Myoporum laetum
Ngaio

Dome-shaped ngaio crowns, with their sprawling, corky-barked trunks are a familiar sight on most New Zealand coasts.

Although ngaio may grow to ten metres tall, on exposed coasts it is often low growing, sprawling over cliff faces and rocks. The bright green, glossy, rather fleshy leaves (to 100 × 30 mm) are dotted with pale oil glands and are covered in a thick waxy cuticle that protects them from salt spray.

From October to January clusters of up to six bell-shaped, purple-dotted flowers (each about 10 mm across), with five wide-spreading lobes, appear in leaf axils and are followed by pink, ovoid, centimetre-long fruits.

Ngaio leaves and fruit are very poisonous, and stock deaths from ngaio poisoning still occur. The Maori rubbed infusions of ngaio leaves on the skin to repel sandflies and mosquitoes, and the inner bark was chewed to relieve toothache, while the bruised and warmed leaves made an effective poultice for septic wounds.

Three Kings, North and South Islands — open coastal sites as well as coastal and lowland forest to south Otago.

MYOPORACEAE

P.N.J.

Olearia angustifolia

Olearia angustifolia

Teteaweka

Nearly a hundred years ago Thomas Kirk, in *The Forest Flora of New Zealand*, enthused about the beauty of this shrub daisy of New Zealand's windswept southern coasts: 'This rare and little-known species is one of the most beautiful flowering plants yet discovered, and is attractive alike from the remarkable character of the foliage, the beauty of the flowers and the delicate perfume it exhales.'

Although originally discovered by Dr David Lyall in 1848, while he was surgeon aboard HMS *Acheron*, teteaweka was not fully described until the publication of Thomas Kirk's work 41 years later.

Olearia angustifolia is now often regarded as the same species as *O. oporina*, and the similar *O. chathamica* is also considered to be a variety of *O. oporina*.

Teteaweka is more common in its inhospitable native habitat than Thomas Kirk realised — it is the main component of coastal forest and scrub on exposed coasts of Stewart Island and South Island from Bluff Hill to Preservation Inlet. Its handsome dark, glossy green, saw-edged leaves cover a tight, dome-shaped canopy. Its leaves are leathery, large (70–150 by 10–20 mm) and covered beneath in a dense felt of white hairs or tomentum, which also coats the stout branches. Flower heads are equally emphatic — borne in clusters of three to ten at the branch tips, on stout, white-woolly stalks up to 80 mm long, they measure about 50 mm across and consist of white ray florets ('petals') surrounding purple or pale yellow central tubular disc florets.

Teteaweka grows into a stout shrub or small tree up to six metres tall, with a trunk up to 300 mm through, which is clad in light-brown fibrous bark. Like *Brachyglottis stewartiae*, the trunk often topples over (probably owing to a combination of weakened soil structure through petrel burrowing and frequent gales), and then anchors itself in this new prostrate position by sending down adventitious roots.

South and Stewart Islands — windswept coasts and scrub of Stewart Island and southern and south-western South Island from Bluff Hill to Preservation Inlet.

COMPOSITAE

Olearia angustifolia P.N.J.

A. & S.P.

Plagianthus divaricatus

Plagianthus divaricatus

Saltmarsh ribbonwood

Fragrance rather than showiness advertises the flowers of this relative of the ribbonwood (*Plagianthus betulinus*) and lacebarks (*Hoheria* species). A small (5 mm across), cream, sweetly scented flower accompanies each leaf in early summer, maturing by late summer into a pale buff-coated fruit (about 5 mm across), which usually contains just one hard brown seed.

Thickets of saltmarsh ribbonwood margin estuaries and coastal salt swamps throughout New Zealand, their distinctive highly branched, dome-like shape emphasised by their mauve, sparsely leafy, interlacing branchlets.

North, South, Stewart and Chatham Islands — coastal saltmarshes.

MALVACEAE

C.E.B.

Samolus repens

Samolus repens
Maakoako

This creeping herb forms extensive mats in coastal saltmarshes and rocky crevices reached by salt spray. Tough dark stems (up to 400mm long) produce spoon-shaped, fleshy, olive-green leaves (to 15 by 6mm) and dainty (6mm diameter), white, symmetrical flowers. *Samolus repens* is our only native member of the primrose family.
Throughout New Zealand from the Kermadec to Auckland Islands — in coastal turf, shrubland, saltmarsh and shoreline rocks. Also Tasmania, mainland Australia, New Caledonia and Easter Island.

J.B.

Samolus repens

PRIMULACEAE

J.F.

Selliera radicans

Selliera radicans
Selliera

The 'cut-in-half' white flowers of this mat-forming herb are a distinctive feature near salt water. Fleshy, club-shaped leaves (to 30 by 5mm) form a close sward, which in summer is dotted with white flowers that appear lopsided because of their split corollas.
North, South and Stewart Islands — coastal sands, mud and rock near salt water, and occasionally inland on wet, salty ground. Also south-east Australia, Tasmania and Chile.

GOODENIACEAE

B.G.

Wahlenbergia trichogyna

Wahlenbergia trichogyna

This harebell grows wild only on grass- and bracken-covered cliffs near Blenheim on the north-eastern Marlborough coast. The same species occurs in Australia and may possibly have been accidentally introduced here last century when Australian sheep were landed in Marlborough.

It is the showiest of our ten harebells, producing numerous bluish-violet bell-shaped flowers (each about 30 mm across) on long, slender stalks.
South Island — coastal cliffs among grasses and bracken in north-eastern Marlborough.

FOREST

FOREST

Forest covered most of New Zealand when the Maori arrived about a thousand years ago. There is evidence for fires having occurred naturally before that time but an increase in their extent and frequency since man's arrival has caused the replacement of large areas of forest by shrubland and grassland. In the eastern South Island in particular, changes towards a drier climate have helped prevent re-establishment of the former forests. Evidence of these can be seen in surface logs, hollows left by wind-thrown trees, and from buried charcoals.

Even when European settlement began some 150 years ago, over 50 per cent of the total land area remained in forest, compared with barely 20 per cent today.

New Zealand's forests are usually mixtures of conifers* and hardwoods or southern beeches and hardwoods, or all three. Within these broad groups that make up forests, the proportions of particular species vary with differences in climate, altitude, latitude, aspect, topography and soils.

Lowland conifer-hardwood forests, especially those of the North Island and northern and western South Island, are probably the type most widely regarded as typical New Zealand 'bush'. They are denser and contain more species than other types, particularly than those at higher altitudes and in dry eastern districts. Conifer-hardwood forests also have the multi-layered structure typical of tropical rain forests. The tallest trees in these layers are conifers which emerge above a canopy of hardwoods. Below there is a subcanopy where smaller trees and shrubs such as mahoe (*Melicytus ramiflorus*), marbleleaf (*Carpodetus serratus*), broadleaf (*Griselinia littoralis*), tree fuchsia (*Fuchsia excorticata*), lancewood (*Pseudopanax crassifolius*), pate (*Schefflera digitata*) and mapou (*Myrsine australis*) are prominent. Nikau palms (*Rhopalostylis sapida*) and tree ferns (*Alsophila* and *Dicksonia*) also in this understorey add to the forest's tropical appearance. Lianes abound;

D.H.

Metrosideros umbellata.

*Conifers (from *konos*, a cone; *ferre*, to bear) are also referred to as softwoods, and include trees such as kauri (*Agathis*), rimu (*Dacrydium*), matai and miro (*Prumnopitys*), totara (*Podocarpus*), kahikatea (*Dacrycarpus*), celery pines (*Phyllocladus*) and native cedars (*Libocedrus*).

Southern beeches (*Nothofagus*) include four species (silver, red, hard, and black) and one variety (mountain). Although hardwoods in the strict sense, southern beeches are so distinct and important in New Zealand forests that they are recognised as a separate group.

Hardwoods compared with coniferous softwoods have flowers, not cones. Ratas (*Metrosideros*), kamahi (*Weinmannia*), tawa and taraire (*Beilschmiedia*) are the most important species. Hardwoods are sometimes called broadleaves, in contrast to the needle leaves of many northern hemisphere conifers in the pine family.

G.R.R.

Temperate lowland forest, Fox River valley, West Coast, South Island.

R.M.

Sophora microphylla flowers.

J.B.

Clematis forsteri.

the ropy stems of kiekie (*Freycinetia baueriana* spp. *banksii*), lawyers (*Rubus*), *Clematis*, climbing rata (*Metrosideros*), passion vine (*Tetrapathaea tetrandra*) and native jasmines (*Parsonsia*), twine around and hang from many forest trees. Often only their stems and a few leaves are visible to us — flowers and most of the foliage are higher up where the light is brighter.

The forest floor, however, may have tell-tale signs of tree-top flowers. You may have noticed white clematis petals and the red or white stamens of rata vines along bush tracks during spring or summer. Tree flowers too are often in the canopy far out of sight and reach, although their presence is sometimes revealed by those that have fallen.

Flowers of a few species are more easily visible because they arise directly from the trunk or branches. Typically a feature of tropical plants, this may help make flowers of densely leafy trees more easily accessible to bird and insect pollinators. In New Zealand forests, flowers and inflorescences of mahoe, tree fuchsia, kohekohe (*Dysoxylum spectabile*), mapou and toru (*Myrsine*) and horopito (*Pseudowintera*) arise in this fashion, which is known technically as cauliflory (from *caulis*, a stalk).

Numerous epiphytes perched on the branches and crotches of a variety of host trees also add considerably to the tropical appearance of conifer-hardwood forests, especially those of our warmer and/or wetter regions. Most prominent are the lily family herbs, *Astelia* and *Collospermum*, with their tufts of rather flax-like leaves, which in *Astelia solandri* may be up to two metres long. Ferns, club mosses, mosses, liverworts and lichens are ubiquitous on tree trunks and branches. The grassy-leaved epiphytic orchids *Earina* and *Dendrobium* plainly reveal their true identity only when their fragrant cream flowers appear.

The versatility of some of our native plants in exploiting widely differing niches is exemplified by trees and shrubs that can grow either as epiphytes or on the ground. Puka (*Griselinia lucida*), its smaller-leaved cousin, broadleaf (*G. littoralis*), cabbage tree (*Cordyline australis*) and rata (*Metrosideros robusta*) can all do this. It is interesting that rata, which so often hosts a diverse community of epiphytes, itself usually begins life as one.

The forest floor and fallen tree trunks are luxuriantly carpeted with ferns, mosses, liverworts and lichens, as well as with larger erect herbs such as the forest oat grass (*Ehrharta diplax* = *Microlaena avenacea*), bush flax (*Astelia fragrans*), cutty grasses (*Carex*) and hooked sedges (*Uncinia*), as well as with small, prostrate ones such as pennyworts (*Hydrocotyle*) and *Nertera*.

Forests dominated by one or more of the four native southern beech species (often together with lesser numbers of hardwoods and conifers), usually lack the density and variety seen in conifer-hardwood forests. In the drier South Island mountains particularly, beech forests have a relatively open interior with widely spaced, spindly understorey trees, few epiphytes and lianes, and a sparse ground cover of ferns, lichens and bryophytes.

Some forests consist largely of only one group, i.e. conifers or southern beeches or hardwoods. There are, for example, mountain cedar forests of North Island uplands, kahikatea swamp forests, and mountain beech forests of South Island rainshadow hill country and southern rata-kamahi forests in parts of Westland and south-east Otago.

The forest environment is far from uniform: in lighter, more open habitats along margins (including gaps and clearings) one finds a number of species that quickly colonise open areas but which cannot survive in the shady interior. Many of these also occur in scrub and shrubland, as well as in forest which has been disturbed by slips, roading or logging. Several have most attractive flowers: poroporos (*Solanum*), like so many of their tropical cousins, give a colourful touch with mauve or purple. Wineberry (*Aristotelia serrata*) segregates its sexes on separate trees, although males and females look similar with their pink to deep maroon, incised petals. Tree fuchsia's nectar-rich green and purple flowers are of two types — some larger with long stamens and abundant blue pollen, others smaller, with short stamens and often lacking pollen. Tree tutu (*Coriaria arborea*) is common on both forest margins and in scrub.

One species that reaches a weed-like abundance along some highly disturbed forest margins is pohuehue (*Muehlenbeckia australis*), a light-requiring opportunist that sometimes causes the collapse of tree canopies under the weight of its leafy vines.

Almost all our native trees and shrubs are evergreen, with deciduousness confined to species such as wineberries, fuchsias, lacebarks, ribbonwoods and kowhais. These differ in their degree of winter leaf loss — it is usually complete in fuchsias and ribbonwoods, partial in wineberries and lacebarks, and highly variable in kowhais.

Despite the above-mentioned similarities with the tropics, even our most lush forests lack the prodigious diversity of species and the large, elaborate, colourful flowers characteristic of tropical rain forests. For example, the total number of indigenous flowering plant species in New Zealand is about 1800 — less than half that of tropical New Caledonia, an island about equal in area to the Auckland province.

Apart from a few species with large and/or colourful flowers e.g. kowhais, kaka beak (*Clianthus puniceus*), pohutukawa (*Metrosideros excelsa*) and puawananga (*Clematis paniculata*), the majority of forest plants have small, pale flowers with a simple, radially symmetrical structure. It is thought that this is because New Zealand has only unspecialised pollinating insects such as beetles, flies, moths and short-tongued bees. Such insects only forage from and pollinate simple flowers whose pollen and nectar are easily accessible. They are also indiscriminate, moving freely from species to species. Usually, therefore, our flowering plants do not rely on one pollinator, but are visited by a range of insects, and occasionally also by birds, or even bats and lizards.

J.B.

Hoheria populnea.

J.B.

Pittosporum eugenioides.

C.P.

Dendrobium cunninghamii.

J.B.

The graceful and melodic visits of honey-eaters (e.g. tuis, bellbirds and silvereyes) to the nectar-rich flowers of kowhai, fuchsia, rata and rewarewa (*Knightia excelsa*) will be familiar to many. Birds see a similar range of colours to humans but are especially sensitive to the longer wavelengths of light, so that red, orange and yellow flowers tend to be bird-pollinated. In New Zealand, about 20 of the 30 mainly forest species which are known to be pollinated by birds have flowers of these colours. However, the size, structure and colour of bird-pollinated flowers vary greatly — from the tiny, pale, apetalous ones of native olives (*Nestegis*), to the relatively large red inflorescences of rewarewa. Flowers of a few species (rata, pohutukawa, kiekie, rewarewa and *Collospermum*) are visited by the short-tailed native bat. Although these flowers appear very different, they all have readily accessible pollen and nectar, and are grouped into large conspicuous inflorescences.

For some forest plants — the grasses, sedges and coprosmas — wind is the main carrier of pollen. The stigmas of their flowers are accordingly large and feathery, serving to increase the chances of pollen interception.

Although the majority of forest species flower during spring, summer and autumn, some are notable for extending into or beginning to flower in winter. Winter flowerers include some of the showiest: the terrestrial, narrow-leaved form of kohurangi (*Brachyglottis kirkii*=*Senecio kirkii*) flowers from autumn to early winter, while the beautiful orange-red blossoms of the climbing rata, *Metrosideros fulgens*, appear between February and June.

Populations of kowhai in different parts of the country each flower for about five weeks between autumn and early summer. Some plants such as five-finger (*Pseudopanax arboreus*), three-finger (*Pseudopanax colensoi*), pepper tree (*Pseudowintera colorata*), kanono (*Coprosma grandifolia*) and tree fuchsia begin flowering in winter and often continue into spring and early summer. A few, such as puriri (*Vitex lucens*) and kaka beak (*Clianthus puniceus*), bloom intermittently through most of the year.

A special feature of our forests is their antiquity: relatives of our present-day conifers, hardwoods and southern beeches appeared early in the fossil record of woody plants. Pollens found in ancient rocks show that ancestors of kauri and kahikatea grew in Gondwanaland forests before dinosaurs roamed the earth 190 million years ago. Pollen similar to that of silver beech has been found in rocks 130 million years old. Ancient lineages can also be traced for other forest plants such as rewarewa (*Knightia*), horopito (*Pseudowintera*), rata (*Metrosideros*) and manuka (*Leptospermum*). Isolated in a temperate oceanic climate for more than 80 million years, New Zealand forests have preserved intact, probably more completely than those of any other southern lands, the plants of Gondwanaland.

The following also occur in forest: *Calystegia tuguriorum, Carmichaelia odorata, Cordyline australis, Corokia cotoneaster, Hebe salicifolia, Metrosideros excelsa, Myoporum laetum, Olearia rani, Pratia angulata.*

J.B.

Acianthus fornicatus var. *sinclairii*

Acianthus fornicatus var. sinclairii
Pixie cap

A slender orchid growing from small round tubers, pixie cap is distinguished by its square, red-ribbed flower stem with a single, heart-shaped leaf halfway up it, and by the small, greenish flowers, whose dorsal sepals or hoods have an extended point giving a cap-like appearance — hence the common name. The specific name, *fornicatus*, means 'arched' and refers to the curved profile of the hood. Flowers are each about 5mm across and are borne in groups of two to eight on a stalk about 80mm tall. The lip is concave and reddish. Pixie cap orchids may occur alone or in colonies of up to hundreds of plants.
Kermadec, North, South, Chatham and Stewart Islands — lowland forest and scrub.

ORCHIDACEAE

A.E.

Ackama rosifolia

Ackama rosifolia
Makamaka

First cousin of kamahi (*Weinmannia racemosa*) and towai (*Weinmannia silvicola*), makamaka is a native of kauri forests north of Whangarei, where it often grows together with towai. Makamaka is a small spreading tree up to 12 metres tall with a trunk 600mm through.

The light green compound leaves are very similar to those of towai and consist of six to ten leaflet pairs in juveniles, three to five pairs in adults.

Cream flowers appear between August and November and, although individually tiny (3mm across), they are showy en masse in 150 mm long panicles. By autumn the flowers have matured into small (3mm long), two-valved capsules, whose brownish-red hairs give the plants an overall reddish tinge.
North Island — lowland (especially kauri) forest between Kaitaia and Whangarei.

CUNONIACEAE

J.B.

Alseuosmia macrophylla

Alseuosmia macrophylla
Toropapa

Deliciously scented, trumpet-shaped, deep crimson flowers (up to 40mm long) are the drawcard of this slender shrub. Toropapa is common in northern North Island forests, where it grows in association with such species as kauri (*Agathis australis*), mairehau (*Phebalium nudum*), neinei (*Dracophyllum latifolium*) and kohurangi (*Brachyglottis kirkii*), but is rather rare and local towards its southern limit.

The generic name, *Alseuosmia*, given by Alan Cunningham (1791–1839) to the specimens he collected in the Hokianga in 1826, translates from the Greek as 'perfume of the grove', alluding to the fragrant flowers.

Flower colour varies from greenish white through pink to red, while some plants have white throats with crimson-tipped lobes. The juicy centimetre-long fruits are sweet and edible.

Toropapa grows into a slender shrub about two metres tall, with rather thick elliptic leaves, which vary from 60 to 200mm long and may have entire or serrate margins.
North and South Islands — lowland and montane forest from North Cape south to around Collingwood in the north-western South Island.

ALSEUOSMIACEAE

B.E.

Aristotelia serrata

J.B.

Astelia fragrans

A. & S.P.

Aristotelia serrata

Aristotelia serrata
Wineberry, makomako

Thomas Kirk (1899) suggested that the early settlers gave the name 'wineberry' to this small tree because its berries were used to colour wine. However, petal colour of the small (5 mm across) flowers, which varies from pink to deep red, may equally well explain this common name.

Ten-centimetre-long, branched clusters of flowers elegantly transform wineberry in spring. In colder districts, where it is partly deciduous, the flowers are enhanced by the emergence of fresh-green young leaves.

Wineberry trees usually bear flowers of one sex only. Clusters of 5 mm wide red to black fruits are ripe by midsummer.

Fast growing and tolerant of most soils, wineberry is often seen colonising slips, road cuttings and other open areas.

North, South and Stewart Islands — lowlands to montane forest and forest margins.

ELAEOCARPACEAE

Astelia fragrans
Bush flax, kahaha

Known as bush flax, *Astelia fragrans* has long (to two and a half metres), narrow (40 mm wide) leaves, which are a familiar sight on the floors of lowland and montane forests through most of the country. The leaves, however, differ from those of New Zealand flax in their colour, hairiness and prominent keels. The midrib and two veins flanking the keel are often pale or pinkish, and leaf undersides are covered in a sheen of fine, silvery scales.

During October and November, short, branched stalks bearing pale (male) flowers or dark green (female) flowers appear in the centre of the plant, well protected by the encircling leaves. Plants bear flowers of one sex only, and by autumn each female carries a fleshy panicle of bright orange berries, each about 8 mm across and sitting on a flat, six-lobed plate. The fleshy fruits contain numerous shiny black seeds and are reported to have been eaten by the Maori.

North, South and Stewart Islands — lowland and montane forest, from Thames south.

LILIACEAE

J.F.

Astelia solandri — female flowers

B.G.

Brachyglottis hectorii

Astelia solandri
Kawharawhara

Kawharawhara is confined to wetter forests in the North Island and western South Island. It grows mainly high up in tree forks and attached to tree trunks, often in large clumps, together with the similar and closely related genus *Collospermum*. Leaves are shorter and narrower than in kahaha — usually one to two metres long and 20 to 35mm wide — but mature leaves do have the same coating of silvery scales.

Metre-long flower stalks hang conspicuously from the tuft of leaves, their final 400mm or so comprising a panicle of small yellow, pink or maroon flowers, which develop into globose, 5mm wide, translucent berries, tinted pale green or yellow. Flowering occurs between October and June and fruit may be seen through most of the year.

North and South Islands — wet lowland forest, throughout the North Island and on the west coast of the South Island to south Westland.

Brachyglottis hectorii (=Senecio hectorii)

This native shrub daisy is as eye-catching when seen flowering in the wild as it is in cultivation — as anyone who has driven over Takaka Hill in early January will confirm. Its showy (400mm wide) corymbs of white-rayed, yellow-centred flower heads — each to 50mm across — are teamed with large (to 300mm long), oval, mid-green, serrate leaves.

Fast growing to about four metres and reasonably hardy, *Brachyglottis hectorii* is partly deciduous in colder districts outside its natural range. Together with *Brachyglottis perdicioides*, it is a parent of the vigorous and attractive garden hybrid *Brachyglottis* 'Alfred Atkinson'.

South Island — forests, forest margins and stream banks from north-west Nelson to northern Westland.

R.M.

Brachyglottis hectorii

Brachyglottis repanda — J.B.

Carpodetus serratus — E. & J.K.

Brachyglottis repanda
Rangiora

Distinctive, large (to 250 by 200mm) leaves, dull green above, white below, with wavy, scalloped margins, make rangiora one of our most easily recognised native shrub daisies.

Rangiora is one of the first shrubs to colonise slips and other open sites, its spreading leafy habit providing shelter for the establishment of other plants. Human settlement has extended its range and it is now naturalised as far south as Stewart Island.

Leaf colour is variable and wild populations often include a few plants with deep coppery-purple tinged leaves. Fragrant, frothy-looking panicles of tiny cream flower heads (each about 5mm across) appear in early spring.

A variety, *fragrans*, with even larger leaves (to 500 by 270mm), which grows around the shores of Cook Strait in both the North and South Islands, is popular as an ornamental.

North and South Islands — coastal and lowland forests, forest margins, stream banks, open places from North Cape to Kaikoura. Naturalised around settlements as far south as Stewart Island.

COMPOSITAE

Carpodetus serratus
Putaputaweta, marbleleaf

The Maori name, putaputaweta, translates as 'full of weta holes', referring to the holes often seen in the trunks of these trees. The holes, although attributed to wetas, are in fact made by the larvae of a large native moth. The English common name alludes to the mottled leaves, the main veins of which are often picked out in a darker green than the rest of the leaf.

The small (10mm across), white flowers appear from November to March and are borne in profuse flat-topped clusters up to 50mm across. They slowly mature into black, top-shaped fleshy fruits, about 5mm in diameter, each encircled centrally by a rim marking the point where the sepals, petals and stamens were attached to the flower.

Putaputaweta grows from a rather divaricating juvenile into a small, spreading tree up to ten metres tall with a trunk of 300mm diameter.

The sappy, white wood is tough but not very durable when exposed to the weather, and so its uses were confined to such things as tool handles.

North, South and Stewart Islands — coastal, lowland and montane forests.

ESCALLONIACEAE

J.B.
Clematis foetida

J.F.
Clematis paniculata

Clematis foetida

Few plants are as inappropriately named as this attractive climber: far from being offensive its flowers have a delicious jasmine-like fragrance, which on still, warm spring days hangs heavily on the air.

Between September and November vines bear generous panicles of these yellow flowers, each of which comprises five to eight pale yellow sepals, whose furry undersides provide a ready diagnostic feature for the species. Like other New Zealand *Clematis* species, plants bear flowers of one sex only and male flowers are larger (to 25mm wide) than the females (to 15mm wide).

Leaves are borne on 30mm long stalks with their three leathery dark green leaflets on shorter stalks (5 to 10mm). Leaf shape is variable: young plants have smaller, more divided leaves than adults, while adult leaves may have blunt or acute tips and toothed or entire margins. A further distinguishing feature of this species is the dense, tawny tomentum clothing branchlets and young unfolding leaves. These hairs disappear at maturity, when vines may be up to six metres tall with stout basal stems 60mm thick.

North and South Islands — margins of lowland forest.

RANUNCULACEAE

Clematis paniculata
Puawananga

Puawananga was sacred to the Maori, its starry white flowers symbolising the coming of spring. Its vigorous evergreen vines bear leathery leaves whose stalks twist tightly around the host's stems, often hoisting puawananga right up to the canopy. The white flowers enliven native forests each spring with their profuse treetop displays.

Vines are usually of one sex only, and in the larger male flowers the filaments of the stamens are tipped with a mauve anther. Female flowers have stamens too, but their pollen sacs are empty. Instead, their hairy styles elongate as the seeds attached at the style bases ripen. Clematis seeds may be carried some distance on the wind, buoyed up on these long, silky, silvery plumes.

North, South and Stewart Islands — widespread in lowland and montane forest from the Three Kings Islands south.

RANUNCULACEAE

R.M.
Clematis paniculata

R.M.
White-flowered form of *Clianthus puniceus*

J.B.
Collospermum hastatum

Clianthus puniceus
Kaka beak, kowhai-ngutu-kaka

But for the ornamental value of its showy red, pink or white flowers, kaka beak might well have been doomed to total extinction. Although plants in the wild are now very scarce (only a few near Lake Waikaremoana are known for certain), kaka beak has been very popular as a garden ornamental for over a hundred years.

Sometimes kaka beak forms a small tree up to five metres or so in the wild but cultivated plants (unless trained up a wall) usually grow into soft woody shrubs about two metres tall.

Flowers are about 60 mm long, with one petal, called the standard, reflexed and pointed away from the two side petals and the long curved keel that encloses the style and stamens.

Pea-like flowers and pinnate leaves (i.e. compound leaves with a row of leaflets along each side of the axis), indicate kaka beak's membership of the pea family of plants — a huge and varied assemblage of some 5500 species worldwide, including 50 New Zealand species such as kowhais and native brooms. Familiar exotic pea family plants include ornamentals such as laburnum, wisteria, pagoda tree and black locust, and the widespread and weedy gorse, brooms and lupins.

North Island — forest near Lake Waikaremoana in the central North Island. Possibly also in a few coastal localities and offshore islands from the Bay of Islands to Tolaga Bay.

LEGUMINOSAE

R.M.

Clianthus puniceus

Collospermum hastatum

This large, tufted herb usually grows epiphytically on trees, as does its cousin *Astelia solandri*. Leaves are up to 1.7 metres long and 70 mm wide, and are arranged in a broad fan from which the branched inflorescence of cream flowers arches out. Plants bear flowers of one sex only, like *Astelia*, and between March and August female plants are laden with panicles of globose (5 mm diameter) translucent berries, which are red when fully ripe.
Three Kings, North and South Islands — coastal and lowland forest south to northern Westland and Marlborough.

LILIACEAE

G.W.

Corybas macranthus

Corybas macranthus
Spider orchid

Recorded on Macquarie Island at 54° 30′ S, *Corybas macranthus* also occurs throughout the three main islands, making it one of our most widespread endemic orchids. Popularly called spider orchids, *Corybas* species have long thin sepals and petals, supposedly reminiscent of spiders' legs.

The flowers have three sepals and three petals with one of each specialised: in the outer whorl, one sepal is modified to form the dorsal sepal which overarches the modified petal or labellum. The labellum is usually broad and flat, and serves as a landing platform for insect pollinators.

Up to 100 mm tall, this spider orchid bears a single, long-stalked, oval, fleshy leaf, which is about 35 mm long and has a very silvery underside. The single, dark red and green flower is borne on a shorter stalk than the leaf and is attractively patterned in green and silver. Most plants flower between September and November but a few may be as late as January.

North, South, Stewart Islands and subantarctic islands — stream banks and the forest floor as well as damp, shady sites in shrubland and grassland.

ORCHIDACEAE

J.B.

Dactylanthus taylorii

Dactylanthus taylorii
Wooden rose

This odd plant is the sole New Zealand member of a tropical family of leafless root parasites. The flowers are the only part of the plant appearing above ground; fragrant male and female flowers are borne on separate plants. The male inflorescence has whitish pollen on fleshy pinkish-brown shoots, while the female is dull red and produces small (2 mm long), hard, red-brown, flask-shaped fruits. Flowering and fruiting occur between January and May.

Sadly, this plant is threatened by collectors who dig up its gall-like, warty rhizome (which is attached to a host's root). The globose mass of parasite and host tissue is boiled, leaving the woody, fluted, flattened end of the host's root, known as a wooden rose.

At least 12 trees and shrubs are hosts to this parasite, which may grow up to 300 mm across.

North Island — second-growth lowland and montane forest and forest margins from the Hokianga Harbour south to Mt Egmont and East Cape.

BALANOPHORACEAE

R.M.

Dendrobium cunninghamii

Dendrobium cunninghamii
Lady's slipper orchid

Although it has about a thousand cousins further north in Asia and the Pacific, *Dendrobium cunninghamii* is the sole member of its genus in New Zealand. It has the largest flowers of New Zealand's six epiphytic orchids — up to 25 mm across.

Often perching on tree crotches, this species is equally at home on fallen tree trunks, rocks and in the open, even on the coast, anchored by a mat of rhizomes and roots. It is one of our larger orchids, its yellow, bamboo-like stems growing up to one metre long. The yellow-green, leathery, stalkless leaves (60 by 3 mm) are broadest near the base and each has a pale brown tubular sheath.

Flowers appear during December and January, singly or in clusters of up to six. Petals and sepals are white, faintly veined, and the lip has two reddish or green side lobes and a yellow-green throat.

North, South, Stewart and Chatham Islands — as an epiphyte on trees, fallen logs or rocks, mainly in coastal and lowland forest.

ORCHIDACEAE

R.M.

Dracophyllum traversii

Dracophyllum traversii A.F.M.

Earina autumnalis

Dracophyllum traversii
Mountain neinei

Named after William Travers (1819–1903), a lawyer, politician, naturalist and plant collector, *Dracophyllum traversii* was, in the opinion of Thomas Cheeseman (1846–1923), author of the *Manual of the New Zealand Flora* (1923): '. . . by far the finest species in the genus'.

Certainly its tufts of gracefully arching and tapering reddish leaves (each up to 600 mm long), flaking terracotta-coloured bark and magnificent candelabrum-like habit, combine to give it a most individual appearance.

Mountain neinei is common on many of the wetter mountains of the north and western South Island from Nelson to south Westland. Usually it grows as a small to medium-sized tree up to 13 metres tall, in mixed forest and subalpine scrub. Occasionally, however, it occurs in pure stands, and those who have climbed the Mount Arthur track from Flora Saddle will recall the spectacular sight of mountain neinei en masse.

The red-petalled flowers, although individually small (3 mm across), are crowded into densely branched, pyramid-shaped panicles up to 300 mm long, which are borne at the branchlet tips.

Botanists now consider that *D. traversii* is not a distinct species from the North Island's *D. pyramidale*. *North and South Islands — lowland to subalpine forest and scrub — in the North Island from Great Barrier Island to the northern Kaimai Ranges — and in the South Island mainly on the wetter mountains of northern Nelson, Westland and Canterbury.*

EPACRIDACEAE

Earina autumnalis
Easter orchid, raupeka

Earina mucronata
Bamboo orchid, peka-a-waka

Apart from New Zealand's two species, the other eight members of the epiphytic genus *Earina* occur in the tropical South Pacific.

At first glance both our earinas look rather like the lady's slipper orchid, *Dendrobium cunninghamii*, but they have narrower, shorter, unbranched stems with the leaves lying more or less in one plane, and larger sprays of flowers.

Although the generic name means 'spring', after the flowering time of the species first described (*Earina mucronata*), *E. autumnalis* flowers during March or April, hence its common name, Easter orchid. Its leaves are shorter (50–100 mm compared with 100–150 mm) and wider (5–8 mm compared with 4–6 mm) than those of *E. mucronata* and are distinctly curved, whereas *E. mucronata* has straight leaves. The tubular leaf sheaths of *E. autumnalis* also lack the black spots common on sheaths of *E. mucronata*.

The flowers of Easter orchid are among the most richly fragrant in the New Zealand flora. Appearing in autumn, they are borne in erect panicles of about 40 waxy, white blooms, each about 13 mm in diameter

J.B.

J.B.

Elaeocarpus dentatus

J.F.

Earina mucronata

and with a yellow-orange base to the lip and an orange centre (rostellum), to which the pollen masses adhere until picked up by a pollinating insect.

Perching on trees, or occasionally banks and rocks, the grassy-leaved bamboo orchid is most conspicuous during spring and early summer (October to December), when it produces long, drooping sprays of about 40 faintly fragrant creamy yellow flowers, each about 10 mm across and with a lobed, pale orange lip.

North, South, Stewart and Chatham Islands — coastal to montane forest.

ORCHIDACEAE

Elaeocarpus dentatus
Hinau

In flower, hinau is transformed from a pleasing but unremarkable broadleaved evergreen tree into one of elegant, almost oriental beauty. In late spring racemes of 10 mm wide, white, crimped and cup-shaped flowers hang conspicuously below the stalks bearing them, in the manner of *Styrax japonica*, Japan's snowbell tree. By autumn these have matured into 20 mm long, oval, purplish drupes, the flesh of which was made into an oily cake by the Maori. Kereru, the native wood pigeon, is very fond of hinau drupes and is an important distributor of the seeds.

Growing from a spindly juvenile, which often has oblong, coarsely toothed leaves (rather like those of rewarewa), hinau matures into a straight-trunked, round-crowned tree up to 20 metres tall. Adult leaves are oblanceolate, 120 mm long and widest (up to 30 mm) near their tips, and are covered on the under surface with white, silky hairs.

Hinau heartwood is dark brown, durable and tough, and in pioneer days was used for fencing as well as general construction where totara was not available. From its tannin-rich bark the Maori obtained a fixative that was used for dyeing garments.

A cousin of the pokaka (*Elaeocarpus hookerianus*) and wineberry (*Aristotelia serrata*), hinau has flowers with a similar structure, and the incised petals of all three species give their flowers a distinctive 'crimped' look.

North and South Islands — lowland and montane forest in the North Island, and in the northern and western South Island, but rather rare in the eastern South Island.

ELAEOCARPACEAE

J.B.

Entelea arborescens

Entelea arborescens
Whau, corkwood

Whau, in the words of Thomas Kirk writing in *The Forest Flora of New Zealand* (1889), presents 'an appearance widely different from that of any other native plant'.

Huge heart-shaped leaves, reminiscent of *Acer davidii*, are usually about 250 mm long, but sometimes grow to more than twice this length, and hang from stiff, stout stalks, up to 200 mm long. Leaf margins are coarsely serrated and sometimes slightly lobed, while the undersides are clothed in fine white hairs.

During October and November, drooping clusters of white flowers appear, each flower about 25 mm across and with ruffle-edged petals surrounding a central ring of yellow stamens.

Its fruits are brown, spiny, globular capsules (about 25 mm across) which remain on the plant long after they open to release numerous dark brown seeds.

Whau grows very fast — up to three metres a year — and may eventually mature into a round-canopied tree of ten metres or so, but few specimens live more than a decade. Called 'corkwood' by the early European settlers, the very open, porous wood was used by the Maori for floats of fishing nets. Whau wood is one of the lightest known, and in fact is less than half the weight of cork.

Although one of our most frost-tender natives, it can thrive outside as far south as Dunedin if given a frost-free, sheltered site.

Thomas Kirk in 1889 also wrote that, 'Owing to the progress of settlement, and the practice of setting fire to the scrub growing on the open land, the whau has become comparatively rare in many northern districts where it was plentiful within the last fifteen years. . .'

As its habitat continues to shrink, populations of whau have become increasingly confined to inaccessible coasts.

Three Kings, North and South Islands — coastal and lowland forest mainly along streams and forest margins. It is local south of Raglan and Tauranga but populations occur as far south as Queen Charlotte Sound and Golden Bay in the northern South Island.

TILIACEAE

Entelea arborescens J.B.

A.F.M.

Freycinetia baueriana ssp. *banksii*

Freycinetia baueriana ssp. *banskii*
Kiekie

Kiekie is the most southerly member of a tropical family of climbers with aerial roots, such as the pandanus vines of Pacific islands.

Securely attached to tree trunks by ropy roots, or forming large tangled clumps on the ground, kiekie is one of the many species giving a tropical touch to our forests.

Drooping tufts of minutely saw-edged, strap-like leaves (to 1.5 metres by 25 mm) crowd the ends of stout, branching woody stems. The fibrous leaves were used in Maori weaving.

From September to December, white- and mauve-bracted flowers appear. Up to 300 mm long, the bracts have thick succulent bases and are edible, being known to the Maori as tawhara.

Stalks up to 400 mm long bear fleshy spikes (70 mm × 5 mm) of tiny male or female flowers. The pulp of the ripe fruit was also eaten by the Maori after removal of the bitter-tasting rough outer skin.

North and South Islands — coastal to montane forest, south to Marlborough in the east and Fiordland in the west.

PANDANACEAE

R.M.

Fuchsia excorticata

Fuchsia excorticata

Tree fuchsia, kotukutuku

Heavy, gnarled trunks and pale brown, papery bark, strips of which peel off and hang from the branches (*excorticata* means 'stripped of bark'), make this one of our most easily recognised native trees. It has the added distinction of being one of the few deciduous native trees, although the silvery-backed leaves are only totally shed during the winter in colder parts of the country.

Tree fuchsia's 30mm long, pendent green and purple flowers are produced singly along the branchlets (sometimes on quite old wood), from July to December. The flowers are unusual on two counts: their pollen is blue rather than yellow, and they vary in the size and lengths of their styles and stamens. The latter feature ensures cross-pollination by bees and honey-eaters such as tuis and bellbirds that visit the nectar-rich flowers.

The juicy berries, known as konini by the Maori, ripen to a very dark purple in summer and were used for jam by European settlers.

North, South, Stewart and Auckland Islands — lowland to montane forest and forest margins.

ONAGRACEAE

J.B.

Gastrodia cunninghamii

Gastrodia cunninghamii

Black orchid, huperei

From November to February, depending on locality, this orchid's stout, brown, leafless stems, half a metre or more tall, bear numerous drooping flowers, whose scent has been variously described by some as unpleasant, and by others as sweet. Flowers are short-stalked, about 15mm long and are creamy within, purplish-brown and mottled with pale raised calluses on their outer surfaces.

Lacking leaves and chlorophyll, black orchids receive their nutrients through a symbiotic association with a fungus that parasitises the roots of forest trees. The orchids store nutrients in stout, starchy underground stems, which may be up to 250 by 50mm, and which were once steamed or roasted by the Maori people as a winter food, known as perei.

North, South, Stewart and Chatham Islands — occurring sporadically in lowland forest and shrubland.

ORCHIDACEAE

G.R.

Hoheria lyallii

G.R.

Hoheria lyallii

Hoheria lyallii
Mountain ribbonwood

This slender, deciduous small tree (up to ten metres tall) is usually found growing in groves on forest margins, in gullies and on stream terraces throughout the South Island mountains and on Mt Egmont. Populations west of the Main Divide have darker green leaves and fewer leaf hairs, and were formerly regarded as a separate species, *H. glabrata*.

The large (to 100 by 50 mm), thin, fresh green leaves are teamed in midsummer with conspicuous clusters of saucer-shaped, white-petalled flowers, each about 40 mm across and with five central bundles of yellow-anthered stamens.

The generic name, *Hoheria*, is taken from the Maori name, houhere, which was used for both *H. lyallii* and *H. populnea*. The English common name, ribbonwood, refers to the lacy inner bark.

North Island (Mt Egmont) and South Island — montane to subalpine forest margins, stream terraces and shrubland.

MALVACEAE

B.E.

Hoheria populnea

Hoheria populnea
Houhere, lacebark

Lacebark grows from a whippy, divaricating juvenile into a graceful, pyramidal tree of ten metres or more, reminiscent of a poplar in shape — as alluded to in the specific name, *populnea*. The lacy inner bark, referred to in the English common name, consists of many layers of bast fibres, which were used by both Maori and European women as trimmings for baskets and hats, and for making twine. Lacebark-trimmed bonnets were a popular fashion at the 1885 Wellington Exhibition and also in Nelson late last century.

Fragrant clusters of star-shaped white flowers (each 25 mm in diameter) appear during late summer and autumn in great profusion. They mature into rather strange little dry fruits, each about 7 mm across and comprising five single-winged seeds arranged around a central axis.

The easily split, compact, even-grained timber was once used for inlaying and roofing shingles, but it quickly fell from favour for shingles when they were found to perish quickly.

North Island — coastal to lowland forest and forest margins from North Cape to the Waikato and Bay of Plenty.

MALVACEAE

B.E.

Ixerba brexioides

Ixerba brexioides
Tawari

An elegant tree up to 15 metres tall, tawari is readily distinguished by its handsome leaves and flowers. Narrow, leathery, blunt-toothed leaves (up to 160 by 40 mm) are borne on short (10 mm) stalks. Panicles of white and green flowers (each flower 25 to 35 mm across) appear between November and January, conspicuous at the branchlet tips. About six months after pollination the leathery, capsular fruits split into five valves, each containing two oblong shiny seeds about 5 mm long.

The specific name means 'resembling *Brexia*', a Madagascan tree, while the generic name is an anagram of *Brexia*.

North Island — densely shaded places in lowland and montane forest of the northern North Island from Hokianga to Kawhia and East Cape.

ESCALLONIACEAE

J.B.

Jovellana sinclairii

Jovellana sinclairii
Native calceolaria

A most appealing herb, the native calceolaria is a rather sprawling plant, up to half a metre tall, which bears panicles of 10 mm wide, white and purple spotted, two-lipped flowers. These strongly suggest its affinities with South America's calceolarias and other plants of the snapdragon family.

North Island — shady sites along both coastal and inland forest margins and streamsides in the East Cape/Hawke Bay region.

SCROPHULARIACEAE

J.B.

Knightia excelsa

G.R.R.

Kunzea ericoides

Knightia excelsa

Rewarewa, New Zealand honeysuckle

Rewarewa's narrow, spire-like crown stands out amongst the rounded canopies of other tall broadleaved trees in our forests. This form and its saw-edged, rigid leaves are reminiscent of Australia's banksias, a protea family genus distantly related to *Knightia*. Saplings, however, with their long (to 300mm), thin leaves are much more frail looking than adults with their very thick leaves — buff tomentose beneath — which are rarely more than half the juvenile length. Trees may be up to 30 metres tall, with trunks more than a metre through.

The rusty-red furry-coated flowers are like those of no other native plant. Although they conform to the flower patterns of other protea family plants such as grevilleas and banksias, there are no petals, but instead a showy calyx, which has one stamen fused onto the back of each of its four sepals.

Flower buds appear around September and begin to open by October. In bud, the four narrow sepals form a tube around the style, but when the flower opens, the sepals peel back, leaving pollen on the style, and end up coiled spirally around the base of the flower. With 30 to 50 flowers densely packed in each 100mm long raceme, the effect is of a floral 'pincushion', the 'pins' being the projecting styles.

Although bees and flies visit the flowers, presumably attracted by nectar exuded by four glands at the base of each style, tuis and bellbirds are believed to be the main pollinators.

Rewarewa's woody fruits are also most attractive; densely clad in short rusty-red hairs, like the flowers, their projecting apices represent the remains of the style. When ripe these capsules split down one side to release numerous winged seeds.

The pale red timber is very handsomely grained and figured, and is still prized for inlay and other ornamental work.

North and South Islands — throughout lowland and montane North Island forests, especially in cutover forest of the central region, and also in the Marlborough Sounds.

PROTEACEAE

Knightia excelsa R.M.

Kunzea ericoides *(=Leptospermum ericoides)*

Kanuka

Taller than its cousin, manuka, kanuka can grow into a 20-metre tree with a fibrous-barked trunk a metre or more thick.

Flowers are smaller than in manuka (5 to 7mm compared with 10 to 15mm across) and occur in clusters rather than singly. Their fruits also differ — kanuka capsules are smaller, less woody and are shed by autumn of the year they form, whereas manuka holds its capsules for a year or more and the seed is gradually released.

Like manuka, and exotics such as gorse and broom, kanuka is an opportunist, a coloniser of cleared land and other open sites, where its seedlings have the ample light that they need to grow. However, it does not have the environmental tolerance of manuka, preferring better-drained, more fertile sites.

North and South Islands — lowland to montane forest and shrubland.

MYRTACEAE

A. & S.P.

Libertia ixioides

Libertia ixioides
Mikoikoi, tukauki

This is a most ornamental native herb, the distinctive colour and habit of mikoikoi making it an emphatic landscape garden subject, particularly when placed in groups.

Panicles of 15 mm wide, three-petalled white flowers, centred with three yellow stamens, appear during October and November. The pear-shaped capsular fruits, each up to 20 mm long, remain on the plants through autumn and winter. They open only gradually, long retaining their tightly packed, bright orange seeds.

Widespread from North Cape to Stewart Island, mikoikoi is found wild along streams, banks and rocks. Plants growing in shade can appear quite different, as the leaves are often longer, dark green and lax. In full sun they are stiff, splayed, golden-orange fans, which may be as short as 150 mm or as long as 800 mm.
North, South and Stewart Islands — lowland to montane streamsides, banks and rocks.

IRIDACEAE

E. & J.K.

Libertia pulchella

Libertia pulchella
Forest iris

Slender rhizomes, lightly anchored by thin, wiry roots, put forth dainty fans of stiff upright leaves, about 80 mm long and 3 mm wide. Flowers are relatively large — up to 15 mm in diameter, their spreading tepals conspicuous on the forest floor. They are borne in sparse panicles of up to a dozen during November and December.
North Island and western South Island — damp, mossy places in lowland forest. Also Tasmania, eastern Australia and New Guinea.

IRIDACEAE

J.B.

Luzuriaga parviflora

Luzuriaga parviflora
Lanternberry, nohi

Branching, wiry, zigzag stems of this slender herb creep along damp mossy places on the forest floor. Oblong pointed leaves (7–27 by 3–6 mm) are widely spaced and arranged in two ranks on either side of the stem. From December to March white flowers (up to 18 mm long) appear singly at shoot tips, to be followed by globose, pointed white berries (about 12 mm long).
North, South and Stewart Islands — damp, mossy places in forest from mountains of the Coromandel Peninsula south to Stewart Island.

PHILESIACEAE

J.B.

Luzuriaga parviflora

J.B.

Melicytus ramiflorus

Melicytus ramiflorus
Mahoe, whiteywood

One of our most abundant and widespread small forest trees, mahoe grows in a multi-leadered fashion forming a round-crowned tree ten metres tall.

Mahoe is easily identified by its pale greyish-white bark. The leaves are elliptic, have finely toothed margins, and vary greatly in size (from 50 to 150mm long by 30 to 50mm wide) with site and exposure. Fallen leaves decompose to delicate skeletons showing the intricate vein pattern.

Male and female flowers appear on separate trees between November and February, occurring in clusters of two to ten on short stalks, which are usually attached directly to the bare older branchlets from which leaves have fallen. A few flowers are borne in leaf axils on younger growth. The pale yellow male flowers are larger and showier than the greenish females and are noticeably fragrant. Masses of small (4mm wide), globose, dark purple berries ripen by autumn.

Kermadec, North, South and Stewart Islands — lowland and montane forest and forest margins.

VIOLACEAE

J.B.

Metrosideros diffusa

Metrosideros diffusa
Climbing rata

Sixty species of *Metrosideros* are spread across the Pacific Ocean from New Guinea in the west to Hawaii in the east, with one outlying species in South Africa. Of New Zealand's 11 *Metrosideros* (all endemic), five are trees, the rest lianes or woody climbers.

Our most widespread climbing rata, *Metrosideros diffusa* is a slender liane that grows up to six metres or more through forest trees. Starting in October (in the north) and January (in the south), clusters of small (10mm long) white to pinkish flowers appear; their showiness, as with other *Metrosideros* species, comes from numerous conspicuous stamens arranged around the rim of the floral disc. In *M. diffusa* stamens are white or pale pink, but in most other ratas and in pohutukawa, they are red or orange.

North, South and Stewart Islands — lowland forest.

MYRTACEAE

R.M.

Metrosideros fulgens

Metrosideros fulgens (=M. scandens)
Akatawhiwhi

Hoisting itself up the tallest forest trees by means of aerial roots arising from its stems, akatawhiwhi is one of our larger lianes, with woody stems up to 100mm in diameter. Thirsty bushmen used to relish the cider-like liquid obtained by slitting the stems.

Showy orange to scarlet-stamened flowers are as large as those of the tree ratas but are produced intermittently rather than in a single, brief display. Individual vines also differ in their flowering times, some starting as early as February, others as late as August, so that at least a few can be found in flower through autumn and winter.

North and South Islands — coastal and lowland forest from the Three Kings Islands south to Martins Bay in the western South Island, and Marlborough in the east.

MYRTACEAE

C.P.
Metrosideros umbellata

A. & S.P.
Muehlenbeckia australis

Metrosideros umbellata
Southern rata

Although called southern rata, *Metrosideros umbellata* also occurs in the North Island, but there it is restricted to mountain forests in the Tararua, Thames and Coromandel Ranges and to Great and Little Barrier Islands. It is much more common in the South Island, where it is widespread from lowland to subalpine forests. Its presence on subantarctic Campbell Island at latitude 52° 22′ S makes it our most southerly tree.

In good flowering years southern rata is one of the showiest native trees, its clusters of bright crimson flowers giving otherwise sombre green forests a splash of colour. Flowering can begin as early as December or as late as March (in the south and at higher altitudes).

A close relative of rata (*Metrosideros robusta*), southern rata differs in its glossy, more pointed leaves, which are conspicuously dotted with oil glands beneath, and in its brighter red flowers and larger, woodier, capsular fruits. And unlike rata, which usually begins life as an epiphyte perched high on another tree, southern rata grows conventionally from the ground to a height of 15 metres or more, although in exposed sites it is often gnarled and stunted.

The heavy, strong, pale red wood was used in the early days of European settlement for general construction and agricultural implements.

Hardy, but rather slow growing, southern rata initially forms a round-headed shrub that eventually grows into a densely branched tree with a very neat, rounded crown.
North and South Islands — lowland to subalpine forest from Thames southwards. Much more common in the South Island, especially west of the Main Divide.

MYRTACEAE

Metrosideros umbellata
D.H.

Muehlenbeckia australis
Pohuehue

A member of the dock and wireweed family, pohuehue has stems with the family's typical tubular papery sheath at each node. Stout (basal stems are up to 100 mm thick), deciduous and very vigorous, pohuehue is often a conspicuous climber along forest edges and in scrub or open, disturbed sites such as logged forest.

In coastal scrub near Nugget Point on the south Otago coast, Leonard Cockayne (1928) noted that bracken, thousand-leaved fern (*Hypolepis rugosula*), mahoe and fuchsia were bound into a 'close mass' by *Muehlenbeckia australis*.

Small, greenish-white flowers are borne in crowded panicles up to 150 by 100 mm. Each flower is about 5 mm wide with three frilly stigmas and a perianth, which may become succulent after pollination, swelling around the shining, triangular black nut.

This is the largest-leaved native *Muehlenbeckia*, with thin, oval adult leaves up to 80 by 30 mm. Juveniles have smaller, three-lobed or fiddle-shaped leaves.
North, South and Stewart Islands — forest margins, scrub and disturbed open sites in coastal to montane forest throughout New Zealand.

POLYGONACEAE

G.R.

Nothofagus solandri var. *cliffortioides*

Nothofagus solandri var. *cliffortioides*
Mountain beech, tawhairauriki

Mountain beech and its close cousin, black beech (*Nothofagus solandri*) have the most conspicuous flowers of the native beeches. These are male flowers, whose bright stamens in good flowering years tinge whole forests red. Flowers appear between October and December, depending on the temperature and altitude of the site — generally the warmer and lower a site, the earlier flowering begins.

The smaller, green female flowers are borne on the same twig as males and are wind pollinated. In years of heavy flowering clouds of pale yellow pollen rise like dust over beech forest. This characteristic of periodic heavy flowering and seeding years also occurs in Northern Hemisphere beeches of the genus *Fagus*. Among our native beeches, mountain beech has the lowest frequency (about one year in ten) of good flowering years, while silver beech has the highest.

Mountain beech is also the smallest and shortest-lived native — it averages 15–21 metres in height and lives a maximum of around 350 years.

North and South Islands — sea level to subalpine forests from the Volcanic Plateau and East Cape to Southland.

FAGACEAE

Olearia arborescens J.B.

Olearia arborescens

The first *Olearia* to flower in spring, its large corymbs of white-rayed flower heads brighten forests and shrublands during October and November. Individual flower heads are up to 15 mm across and comprise 15 to 20 florets. The oval, wavy-margined, toothed leaves are a handsome foil for the flowers — they are about 80mm long, 40mm wide, dark green and glossy above and clad in pale, satiny hairs beneath.

Olearia arborescens grows into a shrub or small tree up to 5 metres tall.

North, South and Stewart Islands — lowland to montane forest, scrub and shrubland, south from the central North Island.

COMPOSITAE

J.B.

Parsonsia heterophylla

Parsonsia heterophylla
New Zealand jasmine, kaihua

Like so many New Zealand native plants, kaihua is a master of disguise, having very different leaf forms in juveniles and adults. This variability is also alluded to in the specific name, *heterophylla*. Seedlings are fragile-looking oddities, with their dark green, variously shaped and lobed leaves. Adult leaves are generally unlobed but may be narrow or broadly elliptic, up to about 70mm long and 5 to 40mm wide, on 10mm long stalks.

The sweetly scented flowers and twining growth habit have led to the name 'New Zealand jasmine', although kaihua is botanically related to oleander (*Nerium*) and periwinkles (*Vinca*), not to jasmines. Panicles of about 8mm long, white to yellow tubular flowers, each with five fused petal lobes, appear from September to March. They are succeeded by distinctive, long (150mm), narrow (5mm), brown, woody fruits, which open to release numerous seeds, each equipped with a float of silky hairs.

Three Kings Islands south to Stewart Island — coastal to lower montane forest.

APOCYNACEAE

J.B.
Pennantia corymbosa

R.M.
Peraxilla colensoi

G.R.
Native mistletoe on mountain beech

Pennantia corymbosa
Kaikomako

Kaikomako's profuse panicles of sweetly fragrant, small white flowers appear from November to February. The panicles are about 80 mm long and their white, hairy branches add to the flowers' showiness. Kaikomako trees produce mainly male or female flowers, though a few flowers are bisexual.

This species grows from a tangled, divaricating juvenile into a slender-trunked, grey-barked adult up to 12 metres tall. Leaves on juveniles are much smaller and more widely spaced than on adults. Adult leaves are glossy, broadly oblong and coarsely serrated or lobed, up to 100 mm long and 40 mm wide, and are borne on short stalks about 10 mm long.

The black, fleshy coated drupes (8 mm long) are a favourite food of the bellbird, korimako, and the Maori name kaikomako means 'food of the bellbird'.

The very hard wood was used by the Maori to make fire: a dry, sharp-pointed kaikomako stick was rubbed on a flat piece of mahoe or pate.
North, South and Stewart Islands — lowland and montane forest from Auckland south. Rare on Stewart Island.

ICACINACEAE

Peraxilla colensoi (=*Elytranthe colensoi*)
Native mistletoe, pirita

The showy red flowers of this Southern Hemisphere mistletoe set southern beech trees ablaze with colour around Christmas. The most conspicuous of our native mistletoes when in flower, *Peraxilla colensoi* is a hairless, bushy shrub up to one metre or more across, which gets the best of both worlds by robbing its host of minerals and water, while its own flat green leaves photosynthesise carbohydrates. The mistletoe seedling enters the host's sap stream by means of a special club-shaped enzyme-producing organ, which dissolves the host's tissues.

Although pirita is most often found on silver beech, other species of southern beech, ratas and pittosporums as well as exotics such as *Pyrus* (pears), *Robinia* (false acacia) and *Rosa* (roses) also play host to it.

However, in a recent study C. Ogle and P. Wilson (1985) have suggested that the fondness of opossums for mistletoe may eventually cause its extinction in some areas.
North and South Islands — parasitic mainly on Nothofagus in the eastern and southern North Island and throughout South Island forests.

LORANTHACEAE

G.W.

Pittosporum eugenioides

A. & S.P.

Pittosporum tenuifolium ssp. *tenuifolium*

Pittosporum eugenioides
Tarata, lemonwood

The tallest of New Zealand's pittosporums, tarata grows from a conical juvenile through a compact, round-crowned adolescence, to an open, stoutly branched, pale-barked mature tree up to 12 metres tall. Its English common name, 'lemonwood', alludes to the lemon-like smell emitted by its leaves when crushed or bruised.

The sweetly fragrant, star-like flowers appear between October and December and are 10 to 15 mm across, with five yellow petals, and are borne in compound umbels 50 mm across. After pollination, the black capsular fruits take at least one year to mature. Capsules usually comprise two valves, which eventually open to release a ball of four or five seeds held together by a sticky gum and enclosed in a papery sheath.

Tarata has long been grown as an ornamental in New Zealand and Britain for its handsome evergreen foliage, fragrant flowers, rapid growth and hardiness. The glossy, elliptic, leathery leaves (50–150 by 25–40 mm) are easily recognised by their prominent yellow midribs and slightly wavy margins. The species name *eugenioides* suggests a similarity with *Eugenia*, an unrelated genus of the myrtle family, whose members often have dark green, glossy leaves.

The resin exuded by lemonwood bark was formerly used by the Maori to perfume titoki oil, or mixed with sow-thistle juice as chewing gum to dispel bad breath.

North and South Islands — lowland and montane forest throughout both islands, but particularly in clearings, along forest margins and in cutover forest that has been left to regenerate.

PITTOSPORACEAE

Pittosporum tenuifolium ssp. *tenuifolium*
Kohuhu

Kohuhu and its many variegated cultivars are among the most widely known and grown native plants.

Appearing singly or in small clusters in leaf axils near branch tips during October and November, the deep maroon fragrant flowers darken as they age to purplish-black. They are similar in shape and size to those of karo (*Pittosporum crassifolium*).

Widespread in forest and scrub throughout the North and South Islands, *P. tenuifolium* comprises at least two distinct subspecies: *tenuifolium* with wavy leaf margins, which occurs throughout the North and South Islands, and *colensoi* with flat margins, which is found mainly in western districts of the North and South Islands and also on Stewart Island.

The leaves (30–70 by 10–20 mm) give off a distinctive resinous aroma when crushed. Bark on the trunk and branchlets is black, and the tree is sometimes known as 'black matipo'.

North and South Islands — coastal to montane forest margins and scrub. Absent from the western South Island.

PITTOSPORACEAE

R.M.
Pratia physaloides

Pratia physaloides

This handsome relative of the lobelias grows into a semi-woody shrub up to a metre tall, with thin, soft, dull green, serrate leaves up to 150mm long and 60mm wide.

During spring and summer, racemes of 10 to 15 or more beautiful two-lipped blue flowers appear. Petal bases are deep blue, fading to pale blue at the tips. Each flower is about 50mm long. By autumn bushes carry deep purple 25mm long berries.
North Island (Northland) and Three Kings Islands — streamsides and forest margins.

LOBELIACEAE

E. & J.K.
Pterostylis banksii

Pterostylis banksii
Common greenhood orchid, tutukiwi

One of the most widespread of our 19 native greenhood orchids, *Pterostylis banksii* occurs through most of the country. It flowers between October and December.

Plants vary greatly in size, most ranging from 100 to 300mm tall, but a few may grow to 500mm.

The orchid's four to six keeled, grass-like leaves are up to 15 mm wide and 250mm long. The flowers, too, are relatively large for the genus; up to 35mm long and pale green in colour, with darker stripes. Tips of petals and sepals are orange-pink.

Pterostylis banksii was known as tutukiwi by the Maori, on account of the hood's long, tapering point — like a kiwi's beak.

The flower structure encourages insect pollination: an insect trapped inside a greenhood flower (after landing on and triggering the lip to flick upwards), can only escape through a channel past the stigma and below the anthers. As it crawls out, it may (if it already carries pollen from visiting another flower of this species) pollinate the flower, and will collect pollen masses as it brushes past the anthers.
North, South, Stewart and Chatham Islands — damp, shady sites in forest and forest margins.

ORCHIDACEAE

J.B.

Rhabdothamnus solandri

Rhopalostylis sapida

R.M.

Rhabdothamnus solandri

Rhabdothamnus solandri
Taurepo, matata

The only native member of the tropical and subtropical plant family Gesneriaceae, which includes such popular hothouse ornamentals as gloxinias and cape primrose, taurepo has flowers that immediately reveal its allegiances. The tubular, two-lipped, five-lobed corolla is fused with the filaments of four stamens, whose anthers meet in an arch above the stigma.

The 25mm long flowers are borne singly, nodding at the ends of slender stalks, and are made the more conspicuous by their attractive colour, which may vary from yellow through orange etched with dark red veins.

A slender branching shrub up to two metres tall, taurepo has grey-green, hairy leaves, which are easily recognised by their rounded teeth and dark main veins — the latter giving the leaves a rather mottled appearance.

North Island — streamsides in coastal and lowland forests throughout the island but local in the southern part of its range.

GESNERIACEAE

Rhopalostylis sapida
Nikau

The world's most southerly palm, nikau is an outlier of a family of plants synonymous with the tropics, and a reminder of the many tropical genera present in our flora. Nikau is mainland New Zealand's only palm (another, *Rhopalostylis baueri* var. *cheesemanii*, is found on the Kermadec Islands) and reaches the Chatham Islands and Banks Peninsula, although it is cultivated outdoors at least as far south as Dunedin.

Usually about 10 metres tall, its single, cylindrical, unbranched trunk is crowned by a tuft of about 14 expanded leaves — reminiscent of a feather duster. Leaves are up to five metres long and comprise about 80 pairs of metre-long leaflets arranged in two ranks on either side of the stout midrib.

The sheathing, cylindrical, overlapping leaf bases form the distinctive bulbous-looking expansion, or crownshaft, at the top of the trunk. Every year up to five leaves fall off, each leaving a broad scar encircling the trunk — so although it is appealing to think of trunk scars as representing one year's growth, this is not necessarily the case.

Inflorescences arise from the leaf bases and emerge soon after a leaf is shed (between October and May), breaking through two boat-shaped,

C.P.

C.P.

Rubus cissoides

pink, protective bracts. The stout, pale inflorescence branches then quickly spread out, displaying hundreds of pinkish flowers arranged in threes (two males surrounding a female). Male flowers of a tree open before the female to ensure cross-pollination, which is mainly carried out by flies.

Fruits take about a year to mature into bright red, ovoid drupes, about 10 mm long. They are eaten by native birds, especially pigeons.

The Maori wrapped food in nikau leaves before cooking it, and leaves were also used for baskets, floor mats and thatching. The sweet, crisp young leaves around the central growing tip of the stem were eaten raw or cooked, as were the young flower clusters.
North, South and Chatham Islands — coastal and lowland forest south to Greymouth in the west and Banks Peninsula and the Chathams in the east.

PALMAE

Rubus cissoides

Bush lawyer, tataramoa

Despite the bad reputation its formidable prickly leaves and stems have earned this vigorous vine, its large (600 mm long) panicles of white-petalled flowers are one undeniably attractive attribute. Male and female flowers are borne on separate plants between September and November and are 15 to 20 mm across.

Common in forest, forest margins and clearings throughout the country, bush lawyer can climb 15 metres or more up trees by means of its stout main stems, which may be 100 mm or more thick. Leaves comprise three to five leaflets, which vary in size in different geographical forms but are usually between 100 and 150 mm long and 30 to 60 mm wide. Leaf midveins and stalks are armed with recurved, stout red prickles. Although blackberry-like in appearance, fruits are much less palatable.

The common name, 'bush lawyer', is generally thought to refer to the ability of this prickly vine to snare the unwary.
Three Kings, North, South and Stewart Islands — lowland to montane forest and forest margins.

ROSACEAE

J.B.

Schefflera digitata

G.R.R.

Solanum laciniatum

Schefflera digitata

Pate, seven-finger

A cousin of five-finger, pate has paler green, thinner and more numerous leaflets (seven to nine compared with five to seven), but is almost as common in forests throughout the country. Its delicate leaves are susceptible to drying, and so it grows in damp, shady sites or where humidity is high. Pate forms a laxly branched, spreading shrub or small tree about eight metres tall, and, also like five-finger, often begins life as an epiphyte.

Although small, the yellowish-green flowers are borne in large (to 400 mm across), drooping branched clusters during February and March. Fruits ripen rapidly and are usually mature by the end of March.

North, South and Stewart Islands — damp sites in lowland and montane forest.

ARALIACEAE

Solanum aviculare, Solanum laciniatum

Poroporo

The Maori name includes *Solanum aviculare* and *S. laciniatum*, both laxly branched and softly woody shrubs growing to about three metres. Although they are alike in habit and foliage, their flowers differentiate: lavender coloured and smaller (35 mm diameter), with the petals arranged in a five-pointed star in *S. aviculare*, rich blue-purple and larger (50 mm across) with frilled petals in *S. laciniatum*.

Berries of *S. aviculare* are smaller and rounder than the egg-shaped, 30 mm long berries of *S. laciniatum*. Other differences include chromosome number, pollen and seeds.

The two species occur in Australia and New Zealand. In this country both occur between Auckland and Marlborough, but *S. aviculare* has a more northerly distribution, while *S. laciniatum* extends to the southern South Island. Its presence on Stewart Island is due to its introduction there by European settlers.

The dark green leaves of both species are highly variable in shape, ranging from narrow and unlobed to broader and deeply lobed — usually up to 300 mm long with individual lobes about 30 mm wide.

The Maori people used poroporo in various ways. Leaves were used as poultices, or boiled with lard to give a healing ointment. The inner skin and juices of the leaves were used to treat scabies, as was the juice of the young berries. Juice from the leaves was mixed with soot and rubbed into tattoo wounds.

More recently, poroporo has been grown commercially for a leaf alkaloid used in manufacturing sex hormones and cortisones.

Kermadec, Three Kings, North, South, Stewart and Chatham Islands — shrubland, coastal and lowland forest margins from Kermadec and Three Kings Islands to the Marlborough Sounds and Chatham Islands (S. aviculare), and from Auckland to Stewart Island (S. laciniatum). Both species also occur in Australia.

SOLANACEAE

R.M.

Sophora microphylla

Sophora microphylla
Kowhai

The kowhai is regarded by many New Zealanders as their national flower, and its nationwide distribution and outstanding beauty well fit it for that distinction.

Racemes of golden-yellow, pea-like flowers — each 40 to 50 mm long — bedeck the trees in early spring. Kowhai flowers have five petals: an outer covering standard, two side petals or wings, and a concave keel of two fused petals in which the style and ten stamens lie. Nectar-rich, they are a major attraction for birds such as tuis, bellbirds and waxeyes, which, along with bees, pollinate them.

Once pollinated, flowers quickly develop into brown, rather woody pods, up to 150 mm long, which are most distinctive with their winged edges and the constrictions where some seeds have failed to develop.

Our two other native kowhais have much more limited distributions. *Sophora tetraptera* is confined to the North Island, from East Cape to the Ruahine Range. It is a beautiful tree about the same size (12 metres) as *S. microphylla*, but is more popular as an ornamental as it lacks a twiggy, divaricating juvenile stage. The third native kowhai, *Sophora prostrata*, occurs only in the eastern South Island foothills from Blenheim to a little south of the Waitaki River. It grows into a dense, tangled shrub about two metres tall. Its leaflets are smaller than those of the tree kowhais, its pods lack wings and the seeds are dark maroon-brown, not yellow.

North, South and Chatham Islands — lowland to montane forest, forest margins, riverbanks and open places.

LEGUMINOSAE

R.M.

Sophora microphylla

J.B.

Thismia rodwayi

Thismia rodwayi

This rare and unusual underground parasite lacks leaves and appears only briefly above ground to flower. Its whitish rhizomes run through forest duff and have associated with them mycorrhizal fungi, which facilitate nutrient uptake from the humus.

The semi-translucent, ephemeral flowers (about 15 mm long) appear singly between November and January and, when light shines through them, look rather like tiny red lanterns. The form and arrangement of the stamens and nectaries suggest that the flowers may be pollinated by insects and small animals.

North Island — forest in the central region near Taupo and Tongariro. Also Tasmania and south-eastern Australia.

BURMANNIACEAE

J.B.

Thismia rodwayi

J.B.

Vitex lucens

G.R.

Weinmannia racemosa

Vitex lucens
Puriri

Distantly related to the teak tree of Burma and other South-East Asian countries, puriri shares with its renowned cousin the distinction of a fine, dense, handsome timber. Last century Thomas Kirk pronounced it 'the strongest and most durable of the New Zealand timbers', and it was extensively used for all sorts of outside construction.

Puriri grows into a broad-crowned tree up to 20 metres tall with a stout trunk one metre or more through. Leaves consist of five stalked leaflets radiating from the top of a square stem. The three inner leaflets are the largest — up to 125 by 50mm.

Puriri's pinkish-red, two-lipped flowers are present for at least eight months of the year — one of the longest flowering periods of any native plant. Flowering usually begins in autumn and continues well into spring, but in the north trees also produce a few flowers over summer. Flowers are often borne in sprays of a dozen or more and each measures 25 to 35mm across.
North Island — coastal and lowland forest from North Cape to Cape Egmont in the west and Mahia Peninsula in the east.

VERBENACEAE

Weinmannia racemosa
Kamahi

Together, kamahi and its close cousin towai (*W. silvicola*) are among the most abundant larger trees in New Zealand indigenous forests, where kamahi may grow to 25 metres tall. However, whereas towai occurs only in mountain forests in the northern North Island, kamahi is common throughout the country from Auckland to Stewart Island.

The presence of juvenile and adult foliage in both kamahi and towai has often led to confused identities; kamahi juveniles resemble towai adults, while towai juveniles are very similar to another relative, makamaka (*Ackama rosifolia*). Young kamahi have trifoliate leaves but those of adults are simple. Adult leaves are broadly elliptic, sharply serrate, and up to 100mm long and 40mm wide on stalks up to 20mm long.

Dense racemes (up to 120mm long) of fragrant, white to pale rose flowers appear mainly during December and January. These mature into dry, brown capsules, which remain on the tree long after they have released their tiny seeds.
North, South and Stewart Islands — lowland to montane forest.

CUNONIACEAE

SCRUB, SHRUBLAND, OPEN PLACES

SCRUB, SHRUBLAND AND OPEN PLACES

Plants included here typically grow in a wide range of situations — river terraces and riverbeds, slips, alluvial fans, cliffs and rock outcrops. The dominating environmental influences on them also vary greatly — sites may be wet or dry, windy or sheltered, sunny or shady, and their soils may be rich or meagre.

Although 'scrub' and 'shrubland' are often used synonymously and frequently merge in nature, it may be helpful to point out their differences: scrub is a dwarfed, dense forest of shrubs and small trees with a continuous canopy, whereas in shrubland the shrubs are further apart, with grasses, herbs or bare ground between them. The plants of these associations are on a more human scale than is forest; we can see at close range the variety of colour, form and texture in their flowers and foliage.

The structure and composition of scrub and shrubland communities show a great variety of patterns. Plant height varies with species, age and site — for example, manuka (*Leptospermum scoparium*) can range from one to six metres tall, matagouri (*Discaria toumatou*) from 300 mm to five metres. Communities may consist almost entirely of a single species, or of complex species assemblages, while the area they occupy ranges from small pockets of less than a hectare to many square kilometres. To add to this complexity, scrub and shrubland often occur mixed with grassland.

Leptospermum scoparium

J.B.

J.B.

How did these communities arise? In many areas scrub and shrubland are secondary, having originated when forest was destroyed by fire, wind, avalanches, slips, flooding, volcanic activity or, more recently, by logging. Sometimes they are actively progressing back to forest by a process of vegetation change known as succession, whereby different associations of species establish as communities but in time create conditions that give other species a competitive edge, so that they are supplanted.

Light-requiring opportunists, like manuka and kanuka, can very quickly colonise open sites left after forest disturbance. Young manuka/kanuka stands are dense but eventually thin out, allowing hardwood species such as karamu (*Coprosma robusta*), kohuhu (*Pittosporum tenuifolium*), mapou (*Myrsine australis*) and lancewood (*Pseudopanax crassifolius*) to establish in gaps. Fruits of some hardwoods are attractive to birds, especially tuis and wood pigeons, which void seeds of other species, including conifers, when feeding, thus spreading a variety of hardwoods and conifers which eventually overtop and suppress the colonist scrub. Manuka is shorter and more rapidly overtopped than kanuka, and so successions through it are more rapid — usually less than 100 years. Those through kanuka may take 200 years or more.

In areas too dry for forest to re-establish, such as arid inland basins of the South Island, scrub and shrubland remain as the stable permanent vegetation. Manuka and kanuka are frequently important again here, as are matagouri, tree tutu (*Coriaria arborea*) and mountain cottonwood (*Cassinia vauvilliersii*) — either alone or in mixed stands. Although they are not legumes, both tree tutu and matagouri have nitrogen-fixing nodules associated with their root systems, which assist their colonisation and survival on infertile stony sites. Native brooms (*Carmichaelia*, *Notospartium* and *Chordospartium*), which frequently grow in dry, windy sites, exhibit both the usual nitrogen-fixing ability of legumes and the summer leaflessness of matagouri.

Manuka and kanuka deserve special mention as the most widespread and abundant plants of lowland and montane scrub. Manuka is in fact the most remarkably adaptable, tolerant and widespread of all our woody native plants — in summer its fragrant, white, myrtle flowers will greet you from as far north as the Three Kings Islands to Stewart Island in the south. Over this range it thrives in habitats as diverse as the meagre soils of Northland's gumlands, the dry pumice of the central North Island's Volcanic Plateau, Westland's pakihi bogs and along wet, peaty lakeshores, tidal rivers and coastal strips in the southern South Island. Nor do high concentrations of sulphur, magnesium or heavy metals (which would be toxic to most other native plants) deter it. Manuka is one of the few species able to grow around the central North Island's thermal pools and in the ultramafic mineral belt of Nelson, Marlborough and western Otago. Its very success has led to its designation as a weed in some areas of infertile hill country, but not even manuka blight (a scale insect introduced from its cousins in Australia) has been able to contain it.

While kanuka may eventually grow to a tree of 10 to 20 metres, and will dominate mixed stands with manuka, it lacks manuka's adaptability and grows only in relatively well-drained sites.

Manuka/kanuka scrub is often rich in species and life forms — from ground orchids (e.g. *Acianthus fornicatus* and *Thelymitra carnea*) to lianes (e.g. lawyer, clematis, pohuehue and native jasmine), epiphytic lichens and even mistletoes (e.g. *Korthalsella salicornioides*).

Like those of most New Zealand flowering plants, flowers of manuka and

R.M.

Limestone cliffs of Woodside Gorge, Marlborough, are home to native broom species. *Notospartium glabrescens*.

Pomaderris kumeraho.

J.B.

E. & J.K.

Thelymitra decora.

kanuka are visited by a range of insects (flies and bees). Less common guests are the nectar-eating geckos (*Naultinus* spp.). The pollinators of other scrub species include beetles (e.g. on *Clematis*), flies, bees, butterflies, moths and birds.

The divaricating habit and small leaves of many scrub and shrubland species givc these communities their characteristic dense grey-brown appearance. One of the best known of these is matagouri, the dominant species of riverbed and river-terrace scrub and dry intermontane basins. The modesty of its small, cream, apetalous flowers is compensated for by their powerfully sweet fragrance. When it grows together with other shrubs, many of matagouri's associates are also divaricates, such as corokia (*Corokia cotoneaster*), mountain wineberry (*Aristotelia fruticosa*), mingimingi (*Coprosma propinqua*), porcupine shrub (*Hymenanthera alpina*), niniao (*Helichrysum aggregatum*) and *Olearia virgata*. Koromiko's (*Hebe salicifolia*) willow-like, bright green leaves and gracefully arching racemes of white flowers enliven the brown, grey and mauve traceries of the divaricates.

Certainly scrub and shrubland have their share of florally outstanding plants: kumarahou (*Pomaderris kumeraho*) of Northland's gumlands flowers so prolifically that its branchlets are often hidden, while the inappropriately named *Clematis foetida* bears among the most deliciously fragrant flowers in the flora.

Communities on dry, open rocky sites are characterised by species of e.g. *Hebe*, *Carmichaelia*, *Pimelia*, *Notospartium*, *Pachystegia*, *Hymenanthera* and *Brachyglottis*. Their forms and flowers range through the leafless, semi-weeping pea-flowered pink broom (*Notospartium carmichaeliae*), the red-margined leaves and mauve flowers of Marlborough lilac (*Hebe hulkeana*) and the leathery-leaved Marlborough rock daisy (*Pachystegia insignis*), whose stout, furry flower stalks bear handsome, white-rayed daisy heads.

The need for conservation is urgent for scrub and shrubland communities, as many adjacent to agricultural land come under increasing pressure — through invasion of their habitat by aggressive exotics such as gorse and broom, or from grazing by stock and wild goats. Even attempts to control gorse and broom by aerial spraying of herbicides have, ironically, further threatened several native broom species and associated shrubs.

The existence of a number of species is made even more precarious by their extremely localised distributions; the beautiful pink-flowered Akaroa daisy (*Celmisia mackaui*), for example, is now confined to just a few rocky sites in mixed scrub and grassland in the south-east corner of Banks Peninsula.

The following also occur in scrub, shrubland and open places: *Acianthus fornicatus* var. *sinclairii*, *Aristotelia serrata*, *Brachyglottis repanda*, *Clematis foetida*, *Corallospartium crassicaule*, *Cortaderia toetoe*, *Corybas macranthus*, *Epilobium glabellum*, *Fuchsia excorticata*, *Gastrodia cunninghamii*, *Helichrysum bellidioides*, *Hibiscus trionum*, *Kunzea ericoides*, *Libertia ixioides*, *Montia australasica*, *Muehlenbeckia australis*, *Olearia arborescens*, *Phormium tenax*, *Pittosporum tenuifolium*, *Pratia angulata*, *Raoulia subsericea*, *Selliera radicans*, *Sophora microphylla*, *Wahlenbergia albomarginata*.

P.N.J.

Calystegia tuguriorum

R.M.

Carmichaelia odorata

Calystegia tuguriorum
Native bindweed

Smaller flowers and leaves distinguish native bindweed from the introduced weed convolvulus (*Calystegia silvatica*). Its creeping, branched rhizome sends up slender twining stems, which bear thin heart-shaped leaves up to 40 mm long.

During early summer showy white flowers appear in the leaf axils of long slender stalks. By autumn these have matured into brittle pale brown capsules, each with a central pointed peak and containing several relatively large (4 mm long), hard, orange seeds.

North, South and Stewart Islands, also the Three Kings and Chathams — scrub, forest margins and forest.

CONVOLVULACEAE

Carmichaelia odorata
Maukoro

Apart from one on Lord Howe Island, the 38 species of *Carmichaelia* occur only in New Zealand. Most native brooms are leafless as adults, with stiff or whippy green photosynthetic branches and small, pinkish, purple-veined pea flowers.

Found throughout New Zealand, *Carmichaelia* species occupy a diverse range of habitats from bogs and streamsides to rocky, semi-desert areas.

Carmichaelia odorata is atypical of its genus in retaining leaves as an adult and it grows in far more equable places (streamsides and forest) than do most native brooms. It forms a semi-weeping, much-branched shrub up to two metres tall, and is covered during spring and summer in small (to 10 mm long), pinnate leaves comprising five to seven leaflets. The green branchlets are about 2 mm wide, flat and finely grooved.

White, purple-veined flowers are borne profusely in spring and mature into straw-coloured pods about 7 mm long.

North Island — lowland and montane forest and streamsides of central and southern regions.

LEGUMINOSAE

Cassinia vauvilliersii B.G.

Cassinia vauvilliersii
Mountain cottonwood

Contrary to the implication of its common name, this compact, vigorous shrub daisy is not confined to mountains but occurs widely in shrubland and tussock grassland, from the coast to the subalpine zone.

Its ability to sprout quickly from a basal crown has led to its prevalence in depleted burnt and grazed areas — to the extent that it and other *Cassinia* species have been listed as noxious weeds in some areas. The small (5–12 by 2–3 mm) leaves are green above, dull yellowish felted below, and their margins curve under. Flowers are grouped into narrow (4 mm across) heads, which are in turn arranged in tight, flat-topped clusters of 10 to 20 at branch tips. Flowering occurs mainly between December and February. The heads appear white because the bracts surrounding them have white tips, and narrow white scales are interspersed with the florets. Several varieties (differing in tomentum colour and leaf size) have been recognised, although it is now thought they may not warrant varietal status.

North, South, Stewart and the subantarctic islands — coastal to subalpine shrubland and tussock grassland.

COMPOSITAE

Celmisia mackaui A.E.

Celmisia mackaui
Akaroa daisy

The few remaining wild populations of this handsome Banks Peninsula herb are confined to inaccessible gullies and rock outcrops out of reach of grazing stock. However, with adequate protection in fenced reserves, *Celmisia mackaui* is now re-establishing in snow tussock grassland along the steep, rocky summit ridge behind Akaroa.

Like the other three lowland celmisias, (*C. lindsayi*, *C. hookeri*, *C. morganii*), *C. mackaui* is much more amenable to cultivation than its alpine cousins.

Among the large-leaved species it is unusual in having hairless leaf undersides. The sword-shaped leaves are dark matt green above and glaucous below, and may grow to 500 mm or more, forming robust tufts up to 600 mm tall.

Its horticultural appeal is furthered by its purple leaf bases and flower stems, and the showy white, or in some plants, pink ray florets of its large (50 mm wide) flower heads.

South Island — coastal cliffs and rocky places in tussock grassland and forest margins of south-eastern Banks Peninsula.

COMPOSITAE

Chordospartium stevensonii D.G.

Chordospartium stevensonii
Weeping broom

Dense clusters of pale lavender flowers borne on slender pendulous branchlets transform this leafless and otherwise lifeless-looking small tree each summer.

Like other native brooms (refer also to pink broom, *Notospartium carmichaeliae*), weeping broom is under threat of extinction in the wild, owing to aerial herbicide spraying of gorse and broom in inland valleys of eastern Marlborough. It is thought that there may be fewer than 100 wild plants of this species remaining.

Strangely, weeping broom was probably cultivated in England nearly 50 years before it was rediscovered in New Zealand by George Stevenson in 1909. It was formally described in 1911 by the eminent New Zealand botanist Thomas Cheeseman.

It has been speculated that weeping broom arose through hybridisation of pink (*Notospartium*) and coral (*Corallospartium*) brooms.

South Island — eastern Marlborough on alluvial soil along tributaries of the Awatere, Wairau and Clarence Rivers and a few coastal gullies north of Kaikoura.

LEGUMINOSAE

Cordyline australis

Cordyline australis
Cabbage tree, ti kouka

The cabbage tree's fragrant, small, white flowers are borne in huge panicles up to one and a half metres long and 600 mm wide. They mature into white berries 5 mm across, which contain several black, shiny seeds.

The flowers are lily-like in structure and indeed the cabbage tree was once placed in the lily family. However, botanists now classify the larger, woody lily-family plants such as cabbage tree and New Zealand flax in a separate family, which includes Mexico's *Agave* and *Yucca* and Africa's *Dracaena* and *Sansevieria*.

The cabbage tree provided the Maori with food — a sweet porridge was made out of the dried pith and roots, while central buds and young leaf bases were also eaten — and its leaves were woven into sandals and baskets.

Sufficiently widespread to be a symbol of the New Zealand landscape, the cabbage tree grows in many different habitats throughout the country — from coastal cliffs and lowland swamps to bush margins and tussocky hillsides.

North, South and Stewart Islands — lowland to montane forest margins, swamps and open places.

AGAVACEAE

Cordyline australis

J.F.

Corokia cotoneaster

D.H.

Gaultheria crassa

Corokia cotoneaster
Corokia

Pioneer botanist Allan Cunningham (1791–1839) coined the generic name *Corokia* from the Maori korokio, which referred in particular to *C. buddleioides*, a species confined to coastal and lowland forest of the northern North Island.

French botanist Etienne Raoul was the first to collect and name (in 1840 on Banks Peninsula) *C. cotoneaster*, the smallest but most widespread of our three native *Corokia* species. The specific name he gave it presumably alludes to the resemblance of the plant in fruit to European shrubs of the genus *Cotoneaster*. In Britain this relatively hardy native shrub is known as the 'wire-netting bush' because of its densely divaricating twiggy form.

The zigzag black branches bear spoon-shaped leaves whose tapered, flattened petioles (up to 20mm long) are usually longer than the leaf blades. There is considerable natural variability in the size, colour and shape of the leaves, and plants range from less than half a metre on exposed sites to pyramidal specimens three metres or more tall where conditions are favourable.

Where *C. cotoneaster* and *C. buddleioides* grow together in the northern North Island, natural hybrids are quite common. These show a range of forms, and a number of selected clones of *C.* x *virgata* are popular in cultivation, for example, 'Cheesemanii', 'Bronze King' and 'Little Prince'.

Bright yellow, 10mm wide clusters of starlike flowers appear either in spring in cooler localities or throughout much of the year in the north.

Three Kings, North and South Islands — widespread from coastal to montane rocky places, river flats, shrubland and forest margins.

CORNACEAE

Gaultheria crassa

Showy racemes of waxy, white, bell-shaped flowers plainly give away the heath family allegiances of this attractive shrub. Its thick rigid leaves often have red petioles and leaf tips. Up to four racemes (about 40mm long) appear at branch ends between October and December and mature into round, dry, brown capsules. By contrast the related snowberry (*Gaultheria depressa*) has fleshy white or red fruits.

North and South Islands — subalpine to low-alpine rocky sites from the Ruahine Range south.

ERICACEAE

R.M.

Hebe cupressoides

R.M.

Hebe diosmifolia

Hebes

Apart from a few species that grow in wet places such as alpine bogs and peaty ground, one feature links nearly all of the other hebe species you will see growing — a preference for rocky and stony places, whether these be on coastal cliffs, river bluffs or mountain outcrops.

Hebes show a remarkable range of size and shape — from the coastal koromiko of wind- and wave-lashed coasts to the tough little 'whipcords' of harsh mountain habitats with their tightly overlapping cypress-like scale leaves. Only 22 of our 80 or so species occur in the North Island, and eight of these are also found in the South Island. The remainder are South Island endemics, among which the mountain dwellers are the most numerous, and show the greatest variety of size, shape and flower arrangement.

Hebe cupressoides

This species is unusual among whipcords on account of its very narrow (1 mm diameter) bluish-tinged shoots and widely separated scale leaves — those of other whipcords are closely overlapping so that the stem is not visible between them.

Now rare in the wild, it occurs sporadically from Marlborough to Otago, and is best known from the Castle Hill area in central Canterbury.

It is a popular species in home gardens because of its attractive colour and form, compact size (one by one metre), hardiness and rapid growth. Groups of six to eight pale mauve flowers are produced at the shoot tips during December and January.

South Island — eastern regions from Marlborough to the Otago lakes on montane to low-alpine river flats and terraces.

SCROPHULARIACEAE

Hebe diosmifolia

In the wild there are two forms of this hebe — a tall type (to six metres) that produces white flowers in summer, and the more widely cultivated shorter (to one metre) form that bears its white or mauve flowers in spring. The branched flower stalks are of different lengths so that the flower clusters form flat-topped corymbs, which project conspicuously above the leaves. The elliptic leaves, with their narrowly incised margins, readily distinguish this species from all other hebes.

North Island — scrub in the northern part of the island.

SCROPHULARIACEAE

B.G.

Hebe divaricata

Hebe divaricata

Despite its impressively profuse flowering, making 'the whole bush a foamy white mass' (L. J. Metcalf, 1972), this most desirable species is rarely seen in the nursery trade or home gardens. And this is not because it is difficult to propagate or grow: it comes away very easily from cuttings (as do most hebes) and is hardy throughout the country, thriving in sun or shade.

The narrowly elliptic leaves (20–30 by 3–4mm) are shining above, but dull below. Branched clusters of white flowers are borne below the shoot tip, crowding together and often almost hiding it.

South Island — rocky river banks along forest margins of Nelson and northern Marlborough.

SCROPHULARIACEAE

Hebe hulkeana R.M.

Hebe hulkeana

Marlborough lilac

Like *Hebe divaricata*, this species is endemic to the north-eastern South Island, occurring on limestone bluffs. In early summer, loose panicles of white to pale pinkish-mauve flowers team with its purplish-red branchlets and shiny red-edged serrate leaves. Populations vary a lot and some plants have more bluish flowers, while others have redder and rounder leaves. Leaves vary in size from 25 to 50mm long by 12 to 25mm wide.

Plants grow in a straggly open-branched manner to about one metre tall, and during spring and early summer (October - December) bear 50 or more terminal 250mm long panicles.

Various forms of this species are now available from nurseries, including a number of hybrids. One of the earliest was a cross between *H. hulkeana* and *H. lavaudiana*, which originated last century in William Martin's nursery at Fairfield near Dunedin, and was called *Hebe* 'Fairfieldensis'.

South Island — lowland to montane rocky bluffs and banks from Marlborough to the Conway River in North Canterbury.

SCROPHULARIACEAE

R.M.

Hebe pimeleoides var. *pimeleoides*

Hebe pimeleoides var. *pimeleoides*

Prostrate to half-upright in habit, this small (to 300mm) shrub is another hebe widely grown as an ornamental.

The specific name means 'resembling *Pimelea*', an unrelated genus of mainly mountain shrubs having a similar habit and leaf colour. Dark purple stems bear small (5–15 by 2–6mm), widely spaced, thick, blue-green leaves, which are often red-margined.

From November to March plants produce short lateral racemes of distinctive bluish-purple flowers.

South Island — stony lowland to montane sites from Marlborough to South Canterbury.

SCROPHULARIACEAE

R.M.

Hebe raoulii var. *raoulii*

C.P.

Hebe salicifolia

R.M.

Hebe raoulii var. *raoulii*

Hebe raoulii var. *raoulii*

Another favourite with rock gardeners, this small (100 to 300 mm tall) shrub bears thick, spatula-shaped leaves with red, incised margins. From as early as September until as late as March, plants produce small (60 mm long) terminal panicles of charming pink to mauve flowers.

This is a variable species and three distinct wild varieties are known.

South Island — dry, rocky lowland to montane sites in Marlborough and Canterbury.

SCROPHULARIACEAE

Hebe salicifolia

Koromiko

The specific name of this shrub means 'willow-leaved', on account of its lanceolate, bright green, rather thin leaves, which were well known to the early settlers as an effective cure for diarrhoea — a herbal tea of leaves was drunk or the bitter-tasting buds chewed.

Flowering occurs in midsummer, the white or pale lilac flowers being borne in slightly drooping racemes 100 mm or more long.

South and Stewart Islands — lowland and montane scrub and forest margins, especially along streamsides and in damp places. Also on the coast of Chile at about the same latitude.

SCROPHULARIACEAE

P.N.J.

Hymenanthera alpina

Hymenanthera alpina
Porcupine shrub

This low-growing shrub's rigid, pointed, grey branchlets have earned it the common name porcupine shrub. In very exposed sites plants have few leaves and hug the ground (or a rock) closely, but with more shelter are taller and leafier.

During December and January small, yellow, bell-shaped flowers are borne on the undersides of branchlets, and the unusual outgrowths joining the five stamens explain the generic name: *hymen*, 'a membrane' and *anthera*, 'anthers'.

Fleshy-coated white and purple fruits containing one or two large black seeds mature by autumn.
South and Stewart Islands — rocky coastal to subalpine dry places in shrubland, grassland and herbfield in the eastern South Island and Stewart Island.

VIOLACEAE

P.N.J.

Hymenanthera alpina

P.N.J.

Hypericum japonicum

Hypericum japonicum
Matted St John's wort

Diminutive compared with its several naturalised cousins such as tutsan (*Hypericum androsaemum*) and St John's wort (*Hypericum perforatum*), matted St John's wort is a herbaceous perennial, whose spreading slender stems form small flat green patches in open rocky places throughout New Zealand.

Small, bright yellow, five-petalled flowers appear in leaf axils during November and December.
North, South and Stewart Islands — open rocky places near water in lowland to montane sites. Also Tasmania, mainland Australia and eastern Asia as far north as Japan.

HYPERICACEAE

J.B.

Lagenifera pumila

Lagenifera pumila
Papataniwhaniwha

The dull green rosettes of this small herb form loose patches growing from branched slender rhizomes. The leaves are thick, conspicuously hairy and usually club shaped, with slender petioles and broad, serrated blades.

During spring and early summer the rosettes produce slender flower stalks up to 160 mm long, each bearing a small (10 mm wide) head of pale yellow, white-rayed flowers.
Kermadec Islands south to Chatham, Stewart and Campbell Islands — coastal to montane open places in grassland and light forest.

COMPOSITAE

R.M.
Leptospermum scoparium

Leptospermum scoparium
Manuka

Distinct from kanuka but often confused with it, manuka is a large shrub or small tree (up to 4 metres tall) with a remarkably wide ecological range. It may be found growing in seaside sands, bogs, dry rock outcrops, or leached gravels from the coast to well into the subalpine zone (1500 metres or more). A light-demanding opportunist like kanuka, manuka is quick to colonise open and disturbed areas, especially after fire, which releases its thin seeds from their woody capsules. However, successions involving manuka usually last less than 100 years, compared with the relatively slow changes within kanuka stands.

Their flowers differ too: those of manuka are larger (up to 15mm across), are borne singly, more sparsely, lack a stalk and stand more erect than kanuka flowers with their spreading stamens.

Flower size and colour as well as leaf characters vary considerably throughout the country: plants in North Auckland and Northland with larger pink flowers and narrower leaves are sometimes recognised as a variety, *incanum*, popularly known as pink manuka. Although wild-derived plants of pink manuka are seldom cultivated, some selected forms are. *Leptospermum* 'Keatleyi' is one such form — discovered by Captain Keatley in the Hokianga district during the early 1920s. Its rose-pink flowers are up to 25mm across and the plant grows into a handsome shrub three metres or more tall. It flowers for two months, often beginning in mid-October, although climate may considerably change flowering time.

North, South, Stewart and Chatham Islands — widespread in coastal to subalpine open areas.

MYRTACEAE

Leptospermum scoparium 'Keatleyi' R.M.

Leptospermum scoparium C.P.

J.F.
Mazus radicans

Mazus radicans

This dwarf herb creeps across damp ground by means of rhizomes, which produce short (50mm), leafy, upright branches at intervals along their length. Under some conditions the plant will form a tight mat. The broadly oval, blunt leaves, each up to 50mm long, are easily recognised by the dark 'blanket stitch' markings around their margins.

The two-lipped, four-stamened flowers immediately ally *Mazus* with genera such as *Euphrasia* (eyebrights), *Ourisia* and *Mimulus* — all members of the snapdragon family. The flowers of *Mazus* are borne on short stalks, either singly or up to three together, and are about 12mm long. Petals are white with a yellow and purple throat. Flowers first appear in November and develop during late summer and early autumn into attractive purplish-pink fruits cradled in a persistent, rather fleshy calyx.

North, South and Stewart Islands — lowland to lower-montane swamp margins and grassland from Rotorua south.

SCROPHULARIACEAE

R.M.

Microtis unifolia

Microtis unifolia
Onion orchid

One of our most common and widespread native orchids, *Microtis unifolia* thrives in places as varied as lawns, wood-chip mulch, footpath cracks, grassland and shrubland. Its common name alludes to the single, rolled, onion-like leaf, through which emerges the tall (sometimes to nearly a metre, but usually 100 to 400 mm) spike-like flower stalk bearing 75 or more small, greenish, close-set flowers.

The ear-like extensions to the column on either side of the anther are referred to in the Greek generic name, which translates as 'small ear'. The dorsal sepal forms a broad, cap-like hood, while the lateral sepals are narrower, curve down and point backwards.

Flowering peaks during the November–January period and results in the production of copious amounts of dust-like seed.

North, South, Stewart and Chatham Islands — lowland open places such as banks, grassland, home gardens and lawns.

ORCHIDACEAE

B.G.

Notospartium carmichaeliae

Notospartium carmichaeliae
Pink broom

Like its cousin the weeping broom, (*Chordospartium stevensonii*), this species is threatened in the wild by aerial herbicide spraying of gorse and broom.

In December, pink broom's slender, drooping, leafless branches bear dense, 50 mm long racemes of pink, purple-veined flowers, each about 8 mm long. These mature into flat, pale pods about 8 mm long, which contain up to 10 olive-green, black-speckled seeds.

Pink broom has been known for more than 130 years: Sir David Munro, an early botanical explorer in the north-eastern South Island, discovered it on Christmas Day 1853.

Although still not widely grown in New Zealand home gardens, pink broom makes an interesting feature shrub, its modest size (3 metres in cultivation) and ease of cultivation fitting it for even small city gardens.

South Island — bluffs and rocky places in catchments of the Wairau and Awatere Rivers of eastern Marlborough.

LEGUMINOSAE

B.G.

Notospartium carmichaelae

J.F.

Olearia rani

B.G.

Pachystegia insignis

Olearia rani
Heketara

Heketara's showy flower panicles are a familiar sight throughout spring and early summer in North Island and northern South Island forests. Each flower head is about 10 mm across and comprises 20 or so florets, evenly divided between outer white-petalled ray and central yellow disc florets.

The coarsely toothed, prominently veined leaves have a similar shape to those of another shrub daisy, rangiora (*Brachyglottis repanda*), but are much smaller (50–150 mm by 50 mm). Like rangiora, heketara's leaf undersides, branchlets, leaf stalks and flower stalks are all covered in short, dense, white to buff hairs known as tomentum.

In the open, heketara forms a round-headed shrub of about three metres, but in forests it grows into a small tree of seven metres or more with a trunk up to 300 mm through.

North and South Islands — scrub, forest margins and lowland forest from North Cape to northern Nelson and Marlborough.

COMPOSITAE

Pachystegia insignis
Marlborough rock daisy

One of the aristocrats of the daisy family, *Pachystegia insignis* has foliage as eye-catching as its flowers; the thick, leathery, oval leaves (70–180 by 50–100 mm) are rich glossy green above, while their undersides are covered in a dense felt of pale buff hairs, which form a conspicuous margin. The plant grows into a stout, spreading shrub one to two metres tall.

Several distinct varieties have recently been described (Molloy and Simpson, 1980), one from the Haldon Hills in Marlborough being particularly striking, with terracotta-coloured hairs on the flower stalks and leaf undersides.

The beautiful yellow-centred, white-rayed flower heads are held in a cup made up of rings of overlapping bracts, which give the plant a pleasing, neat appearance. The flower heads sit atop long (to 300 mm), stout, white-felted stalks.

Although Sir David Monro is often credited with having discovered *Pachystegia insignis* — along the banks of Marlborough's Waihopai River in 1853 — Sir Joseph Hooker, who was the first botanist to describe the plant, gives that distinction to Captain David Rough, the first harbourmaster of the port of Auckland, who collected extensively in the mountains of Nelson and Marlborough.

In its native habitat of rocky places along the Marlborough and North Canterbury coasts, as well as further inland, *Pachystegia insignis* is often found growing with other handsome native shrubs such as *Hebe hulkeana, Brachyglottis monroi, Sophora prostrata, Helichrysum coralloides* and native brooms. Regrettably, wild populations of *Pachystegia* and these associated shrubs are declining, owing to browsing by rabbits, sheep and goats, and the encroachment of introduced plants such as broom, wattles and pines.

South Island — coastal to alpine rocky places, mainly between the Wairau and Clarence Rivers in Marlborough. A few populations also occur in North Canterbury.

COMPOSITAE

E. & J.K.
Parahebe catarractae

J.B.
Pernettya macrostigma

J.B.
Parahebe catarractae

Parahebe catarractae

The specific name means 'of the waterfall' — appropriately indicating the plant's habitat on stream banks and in rock crevices, often within reach of the spray from waterfalls.

A most attractive small shrub, up to 600mm tall, *P. catarractae* has the serrate leaves and branched inflorescence typical of its genus. Flowers are about 10mm wide, the disc-shaped corolla comprising four unequal white petals, each of which has purple insect guide markings on its inner surface. Flowers are borne on 15mm long slender stalks in racemes about 200mm long.

Leaves are very variable in texture, size, shape and colour, and range from 5 to 90mm long by 3 to 25mm wide. Four subspecies are now recognised — two mostly confined to the North Island and two to the South Island.
North, South and Stewart Islands — lowland to subalpine rocky places near streams south from the Coromandel Range.

SCROPHULARIACEAE

Pernettya macrostigma
Snowberry

Wiry and flexuous, this low shrub straggles among tussocks and open scrub as well as on peaty and boggy ground. Leaves are narrow (6–10 by 1–3mm), thick, hard, shiny and widely spaced along the stems. Their margins have a few tiny teeth. The small (3mm long), solitary, white to pink, bell-shaped flowers appear between November and January, each in a leaf axil. By autumn they have developed into conspicuous pink or white succulent snowberries, each up to 7mm across and sitting in a fleshy cup made up of the five-lobed, persistent calyx. The style and large stigma of the flowers usually persist in the fruit as well.

Hybrids between this species and the other snowberry, *Gaultheria depressa*, are common, but have broader and more close-set leaves and red or white (not pink) fruits.
North, South and Stewart Islands — montane to low-alpine wet ground in open scrub, tussock and bogs of mountains from Lake Taupo south.

ERICACEAE

R.M.

Phormium cookianum

B.E.

Pimelea longifolia

Phormium cookianum
Mountain flax, wharariki

Despite its common name, *Phormium cookianum* is as common on the coast as it is in subalpine scrub, its other main habitat. Closely related to and sometimes hybridising with New Zealand flax (*Phormium tenax*), *P. cookianum* differs in its pendent, spirally twisted seed pods (those of *P. tenax* are three angled and straight sided), its smaller size and usually drooping (not stiffly erect) leaves.

However, populations of *Phormium cookianum* vary considerably in different localities: North Island plants, which commonly grow on damp cliffs and hillsides up to 800 metres have drooping leaves, bright yellow-green flower tepals and pods up to 220 mm long. Many of the popular variegated flax cultivars were probably selected from this form.

Around Cook Strait and along South Island coasts as far south as Kaikoura in the east and southern Fiordland in the west is an intermediate form, while the third form grows in South Island mountains and Stewart Island, and some of the higher North Island ranges — Tararua, Ruahine and Mt Hikurangi. This last form is more like *P. tenax*, with dark green leaves, often held stiffly erect, red outer flower petals and shorter pods.

Flowers are borne on stout (to 30 mm wide), dark stalks (known as korari or claddy), which arise from leaf bases and grow rapidly during spring to about two metres long. Fifty or more side branches on the upper half of a stalk each bear up to 50 flowers. Like those of *P. tenax*, the flowers are nectar-rich and attract insects and honey-eating birds, especially tuis, bellbirds and silvereyes, which help to pollinate them. The flowers open between November and January and are up to 40 mm long, the three outer tepals varying from yellowish-green to red, while the three inner tepals are greenish.

Phormium flowers, like those of the cabbage tree genus, Cordyline, are lily-like in structure, and for a long time both genera were classified in the lily family of plants, Liliaceae. Now they have been regrouped in the family Agavaceae, which includes the century plant (*Agave*), aloes, yuccas and *Dracaena*.

North, South and Stewart Islands — coastal to subalpine scrub throughout. Especially common on shady slopes.

AGAVACEAE

Pimelea longifolia
Taranga

A most appealing shrub when in flower between September and January, taranga bears compact, terminal globose clusters (to 50 mm wide) of white or rose-flushed, fragrant, tubular flowers. As with our other native pimeleas, the corolla tube of petals is hairy outside.

A larger plant than *Pimelea gnidia*, with a more open branched, leggy form, *taranga* grows to two metres tall.

North and South Islands — lowland to montane shrubland from the Bay of Islands to northern and western South Island. Absent from eastern South Island.

THYMELAEACEAE

J.B.

Pomaderris kumeraho

Pomaderris kumeraho

Kumarahou, gumdigger's soap, golden tainui

The best known and showiest of our eight native *Pomaderris* species, kumarahou is at home on the meagre soils of Northland and Auckland gumlands, where it grows in association with manuka, bracken, towai (*Weinmannia silvicola*), mingimingi (*Cyathodes fasciculata*) and akepiro (*Olearia furfuracea*). The name gumdigger's soap arose because when rubbed between wet hands the flowers release saponins which create a lather.

Usually kumarahou is a shrub two to three metres tall, but in shady or sheltered sites it sometimes grows into a small slender-trunked tree, up to five metres tall.

Flower buds form in autumn, developing slowly over winter to burst forth in early spring into conspicuous, compact, round-headed corymbs of bright yellow flowers. Though individually small (8 mm across), the flowers are borne so prolifically that the branch tips are often completely hidden.

North Island — lowland scrub on poor clay soils from North Cape to the Bay of Plenty.

RHAMNACEAE

Thelymitra longifolia

E. & J.K.

Thelymitra longifolia

Common sun orchid

Ranging from sea level to timber line throughout New Zealand, *Thelymitra longifolia* is one of our most common and widespread native orchids. Flowers of all *Thelymitra* species are unusual in that their sepals and petals are not specialised into the typical orchid lip (labellum) and hood. Flowers of some species only open in full sun — hence the common name. It is thought that this feature may have arisen to protect pollen during wet weather in the absence of a dorsal hood over the column.

Flowers appear between October and February and are borne singly or in groups of up to ten or more. Each is 10 to 15 mm in diameter. The petals are white and sometimes veined reddish-green on their outer surface. At flowering, plants are up to half a metre tall, comprising just the long flower stalk and a single 10 to 20 mm wide, strap-shaped leaf.

North, South, Stewart, Chatham and Auckland Islands — rocks, banks and open ground from sea level to timber line.

ORCHIDACEAE

E. & J.K.

Viola cunninghamii

Viola cunninghamii

Mountain violet

Although the flower structure of our three native violets is plainly similar to cultivated types from the Northern Hemisphere, the petals are predominantly white rather than blue. However, the pale lilac and yellow lines streaking petals of *V. cunninghamii* and *V. lyallii* (and perhaps also ultra-violet lines invisible to us) may serve as guide marks to pollinating insects.

Unlike the two other native violets, which have creeping and rooting stems, the mountain violet grows as single, summer-green tufts from a deep, sometimes woody rootstock. Its oval, crenate leaf blades (10–30 by 10–25 mm) terminate slender, flat petioles (10 to 50 mm long).

Fragile-looking stalks (50 to 100 mm long), which are reflexed at the tip like a shepherd's crook, bear a single flower, 10 to 20 mm wide. Flowers appear over a long period — often beginning in October and sometimes continuing to March. By autumn, 10 mm long, three-valved capsules have developed. These eventually split, flinging out pale seeds.

North, South, Chatham, Stewart Islands — damp, open places from lowland to high alpine sites. Also Tasmania.

VIOLACEAE

WETLANDS

J.B.

Thelymitra venosa.

WETLANDS

Wetland is a term applied to any ground that is permanently or periodically saturated and contains plants tolerant not only of wetness, but of the associated poor aeration. Wetlands occur whenever drainage is impeded, and alongside flowing or static bodies of water. They encompass, for example, lowland swamps, mountain tarns, boggy hillcrests, seepages and flushes on hillsides, river and lake edges, damp terraces and plains, and coastal saltmarshes (the last treated in the section on coastal open places).

Wetland vegetation varies with the interplay of many factors, especially climate, altitude, amount of water flow, availability of nutrients, and aeration. Substrates also range from inorganic gravels or silts to wholly organic peats which result from the accumulation of dead plants that have not decayed. This process is assisted by the acid conditions found in most wetlands. Thus many wetland soils are actively accumulating, and they contain, by virtue of preserved pollen grains and plant fragments, a record of former vegetation at or near the site.

These factors distinguishing one wetland from another are also responsible for variations within a wetland, both in space and time. Thus many extensive wetlands are complexes of intergrading types, while fluctuations in environmental conditions such as the pattern and flow rate of water courses, and the amount of sediment a wetland receives, can promote marked changes in their vegetation over time.

Although swamps and bogs may intergrade, these two wetland types can usually be distinguished by several features: bogs typically form when there is peat accumulation with low fertility, in shallow, slow-flowing or stagnant, poorly aerated water. The reverse generally applies with swamps: fertility is high, the water is deeper, flowing and well aerated.

Lowland swamps are often dominated by raupo (*Typha orientalis*), flax (*Phormium tenax*) or niggerhead (*Carex secta*), or various combinations of these. Giant toetoe grasses of the genus *Cortaderia* are sometimes conspicuous around swamp margins. Under drier conditions, kahikatea (*Dacrycarpus dacrydioides*) and pukatea (*Laurelia novae-zelandiae*) — the latter with aerial roots which assist oxygen uptake — form swamp forest. The cabbage tree (*Cordyline australis*) becomes increasingly important when swamp or swamp forest is drained.

Montane swamps are usually dominated by raupo, although flaxes (*Phormium*), manuka (*Leptospermum scoparium*) and the sedge tussocks *Baumea* and *Schoenus* are also often important.

Bogs are most common in areas with a high rainfall and low summer tem-

P.E.C.

Tarn and bog wetland, south-east Otago.

E. & J.K.

Drosera binata.

peratures — particularly the western and southern South Island, Stewart Island and the subantarctic islands. Lowland bogs of these cooler areas are notable too for their high proportion of mountain species.

Sphagnum moss and the restiad wire rush, *Empodisma minus*, are important in low to high-altitude bogs throughout the country, and are the main peat formers in New Zealand. Both are also very good sponges — sphagnum because of its large, empty leaf cells, wire rush because of its dense-matted rhizomes and aerial roots. Shrubs tolerant of waterlogging, low nutrient levels and acidity include manuka (which may be dwarfed to turf height in some bogs), and species of *Coprosma*, *Dracophyllum*, *Olearia*, *Hebe*, *Cassinia* and *Dacrydium*. Under the most acid conditions, some of these are replaced by the mosses *Sphagnum* and *Dicranoloma*, wire rush, and the tangle fern (*Gleichenia dicarpa*). Red tussock (*Chionochloa rubra*) also tolerates a range of poorly drained conditions, from cold, sour soils to the very wet peat of cushion bogs.

Raised bogs, such as the Kopuatai Peat Dome on the Hauraki Plains, form because peat accumulates faster in the centre of the bog where the water table is higher and aeration and decay are impeded. Raised bogs are characteristic of lowland areas with a moderate and evenly distributed rainfall. In the high rainfall cool uplands of southern and subantarctic New Zealand there are blanket bogs which have a more or less even depth of peat. Also characteristic of these southern and mountain regions are mosaics of species in the genera *Donatia*, *Phyllachne*, *Gaimardia*, *Oreobolus* and *Celmisia*, which form 'cushions' around coffee-coloured peaty pools.

On the West Coast of the South Island are bogs known as pakihis. These have developed on poorly drained glacial outwash gravels which have been leached by the region's high rainfall. They differ from other bogs in having a much thinner layer of peat overlying an impervious iron pan. Tangle fern, wire rush, sphagnum and manuka, together with sedge tussocks, red tussock,

R.M.

Phormium tenax.

Dracophyllum and the fern ally *Lycopodium*, are often abundant on pakihis.

Like so many wetlands, the central North Island's once extensive bogs, dominated by the tall, jointed restiad rush *Sporodanthus traversii*, have virtually disappeared. Now only a few small areas of it remain in the Waikato lowlands and in the Chatham Islands.

Although peat is present in some flushes and seepages, they characteristically have greater decay rates, nutrient levels and aeration than bogs. Both flushes and seepages occupy infilled channels and may be permanent or seasonal, but differ in the types of sites they occupy — flushes occur on valley floors and plains, seepages on hillsides. Their vegetation is very similar, and they are also often seen on lake and river margins. The tussock sedge, *Schoenus pauciflorus*, forms dense red-brown swards 150 to 300 mm high, while most of the associated species are small herbs such as *Pratia angulata*, *Viola cunninghamii*, *Mazus radicans*, *Gunnera prorepens* and *Herpolirion novae-zelandiae*. Sometimes *Schoenus* is absent and a variety of small herbs dominate.

The most florally conspicuous wetland plants are undoubtedly those widely occurring opportunists, flax, cabbage tree, manuka and toetoe. Most wetland plants have smaller, more discreet flowers than these bold performers, yet for variety of structure, colour and arrangement they are second to none.

Orchids such as *Aporostylis bifolia* and *Lyperanthus antarcticus* catch the eye with their precise and perfect sculpture, *Mazus radicans* appeals with its prolific and showy snapdragon-like flowers, while the tiny blue lilies of *Herpolirion novae-zealandiae* can be appreciated for their delicate beauty.

The problem of flowering in a watery environment is overcome in plants such as red pondweed (*Potamogeton cheesemanii*) and bladderwort (*Utricularia monanthos*) by raising flowers on stalks well clear of the water surface. In fact the bright two-lipped purple and yellow flowers of bladderwort are the most obvious part of this tiny specialised plant. Less showy, but en masse equally conspicuous, are the tawny wind-pollinated inflorescences of grasses, sedges and rushes.

As well as being a major wildlife habitat, wetlands are important in flood, drought and erosion control. Functioning rather like a sponge, they absorb flood waters and so help reduce the overflow onto adjacent low-lying land. They also help buffer the effects of drought by releasing water to compensate for low river levels. Erosion and nutrient loss are reduced through wetland plants filtering large amounts of nutrients and sediment.

Many wetlands have a long history of recreational use and some have yielded economically important plants or plant products: peat, flax, raupo and, most recently, sphagnum moss have all been harvested on a cottage or a commercial basis.

Flat and fertile, a great many lowland swamps have been drained and reclaimed: it is estimated that over 90 per cent of New Zealand's lowland wetlands have been significantly modified since European settlement began about 150 years ago. Some have been built on: most of our major cities occupy low-lying former swamps. Reclamation has resulted in the extinction or depletion of several species of plants; a North Auckland orchid, *Thelymitra mathewsii*, is now extinct, while numbers of the marsh fern, *Thelypteris confluens*, and the giant rush, *Sporodanthus traversii*, have greatly declined.

Failure to recognise the importance of wetlands has meant that they are badly under-represented in our reserves and many wetland types of both regional and national importance are as yet unprotected.

The following also occur in wetlands: *Bulbinella hookeri, Celmisia sessiliflora, Cordyline australis, Corybas macranthus, Mazus radicans, Montia australasica, Viola cunninghamii.*

E. & J.K.

Aporostylis bifolia

Aporostylis bifolia
Odd-leaved orchid

Previously known as *Caladenia bifolia* and *Chiloglottis bifolia*, this endemic orchid has now been given a genus all its own, *Aporostylis*. Plants may grow to 250 mm tall, developing from small pear-shaped tubers. A pair of unequal-sized, softly hairy, brown-blotched leaves are borne together at ground level. The longer, wider leaf is 70 to 100 mm long, the smaller about two-thirds this length.

During December and January one or occasionally two small white or pinkish flowers appear at the end of a sturdy stem. Flowers are 15 to 25 mm in diameter and the outer surface is covered in fine hairs.

The dorsal sepal arches hood-like over the column and often opens lemon-yellow but fades later to white with a purple ridge down the middle. Petals are white, apart from a pink median stripe, and the lip has a median ridge of yellow and reddish-brown spots.

North, South, Stewart, Chatham and subantarctic islands — bogs and wet depressions in tussock grassland, herbfield and scrub.

ORCHIDACEAE

G.W.

Caladenia lyallii

Caladenia lyallii
White fingers

From November to February one or two white to pale pink flowers atop a slender stem advertise the presence of this narrow-leaved orchid. At flowering, plants may be up to 200 mm tall and the hairy flower stalk is clasped by one to four bracts (i.e. modified leaves), the lowest conspicuously placed about the centre of the flower stalk.

The dainty flowers are up to 25 mm wide, with one sepal arched hood-like over the lip, which is decorated with four rows of yellowish callii — alluded to in the generic name, *Caladenia*, meaning 'beautiful gland' (from the Greek *kalos*, beautiful, *aden* a gland). The backs of lateral sepals and petals as well as the lip and column have varying amounts of pink or red markings.

North, South, Stewart and Auckland Islands — bogs and peaty areas of tussock grassland, herbfield and open scrub. From sea level (in the south) to the low-alpine zone.

ORCHIDACEAE

E. & J.K.

Cortaderia toetoe

Cortaderia toetoe
Toetoe

Toetoe is our largest native grass (4 to 6 metres tall), whose dense tussocks and graceful, plume-like flower stalks are a familiar site in swamps and damp open ground in the central and southern North Island.

The margins and upper third of leaves (which are up to 3 metres long by 10–30 mm wide) are very rough to the touch.

Botanists now recognise four New Zealand species of *Cortaderia*, each of which is commonly known as 'toetoe'. They can be distinguished by their flowering time, size and habitat.

Cortaderia splendens is the largest (its leaves can be 50 mm wide and its flower stems 6 metres tall) and is confined to the coast and offshore islands between North Cape and Kawhia. It flowers in early December or sometimes earlier on offshore islands. *Cortaderia fulvida* (2–3 metres tall) is smaller and grows throughout the North Island where it is common on stream banks, in forest and on hill country up to about 600 metres. It is the earliest to flower, beginning in mid-November.

The only species native to the South Island is *C. richardii*. Widespread, it extends from the coast to about 600 metres altitude in inland places, and is particularly common on streamsides and other moist ground. Its tussocks

E. & J.K.

Cortaderia toetoe

are 3–4 metres tall although flower stems may reach to 6 metres. It flowers from mid-December until January.

Two South American pampas grasses, *Cortaderia jubata* and *C. selloana*, which were originally introduced for fodder, shelter, erosion control and as ornamentals, have naturalised extensively to the extent where they now constitute a major weed problem from the Bay of Plenty northwards.

Cortaderia toetoe flowers in mid-January and its long (to 1 metre) panicles of tawny flower clusters, called spikelets, terminate stalks up to 6 metres long. Examination of a toetoe spikelet through a handlens reveals two basal chaffy bracts (glumes) and above them two flowers, each tightly invested by two bracts, the inner, the palea, the outer, the lemma. If these bracts are removed, the three stamens and two feathery stigmas of each flower should be clearly visible. Flowers of all grasses have this basic plan, although the number of flowers in a spikelet and arrangement of spikelets vary enormously.

North Island — swamps, damp places and coastal sand hills from Lake Taupo to Wellington.

J.F.

Donatia novae-zelandiae

Donatia novae-zelandiae

This is one of our cushion plants — popularly so named on account of their growth habit. *Donatia* cushions are hard, dark green, flat or gently domed, and up to one metre or more across. Their short (to 100mm long), erect branches are covered in stubby (5–10mm by 1mm), stiff, closely overlapping leaves, which taper to a blunt tip.

Between December and March, flowers are conspicuous on the cushions. Each is about 10mm across and has five rather fleshy, triangular white petals and two short stamens.

Although there are only two species of *Donatia* worldwide (New Zealand's single species and one other in South America), they are considered sufficiently distinct to warrant their own family, Donatiaceae.

The similar native alpine cushion genus, *Phyllachne*, differs from *Donatia* in its very blunt peg-like leaves, which have hairy bases and apical pores, and its flowers with rounded petals and protruding, purple stamens.

South and Stewart Islands — montane (though sometimes sea level in the far south) to subalpine bogs and herbfield. Also Tasmania.

J.F.

Donatia novae-zelandiae

J.B.
Drosera spathulata

P.S.
Drosera stenopetala

Drosera spathulata

On dark, peaty ground this diminutive sundew's rosettes of gland-tipped leaves appear as dainty red circles, some smaller than a one-cent piece. Plants bear one to several flower stalks, each with four or more small white flowers.
North, South and Stewart Islands — lowland to low alpine bogs, and peaty sites.

DROSERACEAE

Drosera spathulata J.B.

Drosera stenopetala
Sundew

This native sundew is easily recognised by its paddle-shaped leaves (up to 70 mm long) tipped with conspicuous stalked, glandular hairs.

From November, each plant bears a single dainty five-petalled white flower about 9 mm across, which sits atop a slender, reddish stalk up to 200 mm long.

The ability of sundews to trap and digest small insects by means of their sticky hairs and enzyme-secreting leaf glands is thought to be related to the hungry soils where these plants grow: the nutrients obtained from digesting the insects supplement the poor supply of soil nutrients.
North, South, Stewart, Auckland and Campbell Islands — montane to subalpine bogs.

DROSERACEAE

E. & J.K.

Lyperanthus antarcticus

Lyperanthus antarcticus

Yellow beaks, horizontal orchid

This robust orchid is usually found in colonies on boggy ground in subalpine areas. At flowering plants are up to 250 mm tall and bear one to three yellow-green flowers (up to 20 mm long), which are partly hidden by the large sheathing bracts below them. The wide, beak-like dorsal sepal is yellow-green, with purplish-brown markings and folds down over the lip, so that the flower appears half closed. The broad lip is often marked with reddish blotches. A pair of rather fleshy leaves sheathe the stem base. *North, South, Stewart, Auckland and Campbell Islands — wet, peaty areas mainly in subalpine bogs and scrub from the Tararua Ranges south.*

ORCHIDACEAE

Phormium tenax

J.B.

Phormium tenax

Harakeke, New Zealand flax

The only feature New Zealand flax shares with linen or linseed flax (*Linum usitatissimum*), a slender, blue-flowered European annual, is that both species yield commercially useful fibre.

Fibre production from *Phormium tenax* was once a major industry in New Zealand, the main use being for making rope. The Maori people had used *Phormium* fibres for centuries before this.

It is a very robust tufted plant, with stiffly erect, dark green leaves up to three metres in length. Emerging from the leaf bases and overtopping the leaves are stout (to 30 mm wide), branched, black stalks, which from November to January bear 700 or more handsome red-petalled, nectar-rich flowers, each up to 50 mm long.

Chiefly pollinated by nectar feeders such as tuis, bellbirds and silvereyes, by autumn the flowers have matured into three-sided black capsules (each about 90 mm long) containing numerous shiny, black, flat oval seeds (each 10 mm long). Whereas its close cousin mountain flax (*Phormium cookianum*) has capsules that are pendent and spirally twisted, *P. tenax* has erect, straight-sided capsules.

Although predominantly a plant of wet ground, *Phormium tenax* is tough and adaptable and occurs in a wide variety of habitats including quite dry sites. Its high natural variability has been the raw material for the selection of numerous variegated ornamental forms popular in horticulture. *Norfolk, North, South, Stewart, Chatham and Auckland Islands — widespread and abundant throughout, especially in lowland swamps, along riverbanks, seepages and other wet ground.*

AGAVACEAE

J.B.

R.M.

Phormium tenax

E. & J.K.

Prasophyllum colensoi

Prasophyllum colensoi
Leek orchid

The common name refers to the single, hollow, tubular leaf from which the flowering stem emerges about one-third the way from the base. Flowering plants range from 100 to 250mm in height. Between November and January each purplish brown flower stem bears five to 20 small, closely set, fleshy red and yellowish-green flowers. Flowers of prasophyllums are 'upside down' because the gynoecium is not twisted during its development as it is in most orchids, and so the flower has its lateral sepals and leaves uppermost. *North, South, Stewart, Chatham, Auckland and Antipodes Islands — swamps, bogs and wet ground in tussock grassland, herbfield and moraine from lowland to low-alpine habitats.*

ORCHIDACEAE

J.B.

Pratia angulata

Pratia angulata
Panakenake

Easily recognised by its rounded, shallowly toothed leaves and white-petalled, lopsided-looking flowers, *Pratia angulata* is common in damp ground throughout the three main islands of New Zealand. Creeping stems are anchored by roots put out at nodes, producing a mat-like growth habit. Leaves vary in size from 3 to 12mm long by 1 to 8mm wide, are more or less round, and often have purplish undersides.

The split between the two narrower petals gives flowers their distinctive asymmetrical appearance. Flowers mature into long-stalked, purplish pink berries, each up to 12mm long. *North, South and Stewart Islands — lowland to subalpine damp places in grassland, herbfield and forest margins.*

LOBELIACEAE

J.B.

Pratia angulata

C.E.B.

Thelymitra venosa

C.E.B.

Utricularia monanthos

Thelymitra venosa
Striped sun orchid

The first New Zealand orchid genus to be discovered and described, *Thelymitra* includes a dozen native species, about half of which are endemic, the rest being shared with Australia. They are commonly known as sun orchids because they frequent sunny sites and the flowers of many species open only when the sun is shining.

Thelymitras differ from all other native orchids in having similar sepals and petals, so that none is specialised as a hood, and the lip is very like other petals.

The specific name, *venosa*, means 'veined' in reference to the sepals and petals, which are striped with deep purplish blue. Although most plants have deep blue flowers, white or pink forms sometimes occur. Flower-bearing stems are up to 500mm tall and between December and February bear three or four (and occasionally up to six) flowers, each up to 20mm wide.

North, South, Stewart, Chatham and Auckland Islands — bogs, and other wet sites from sea level to tree-line. Also Tasmania and eastern Australia.

ORCHIDACEAE

Utricularia monanthos
Bladderwort

This small, fragile inhabitant of peaty bogs and tarns supplements its meagre mineral intake from this nutrient-poor habitat by trapping and digesting minute aquatic animals in its tiny white bladders (each only 1 to 2mm in diameter).

Purple, yellow-eyed, two-lipped flowers are its only conspicuous feature. These appear between November and March and are borne singly, or in pairs, atop slender stalks 20 to 50mm long.

Bladderworts often grow in loose mats, with their narrow summer-green leaves (5 to 20mm long) attached to slender, creeping white stems.

North, South and Stewart Islands (more widespread in the South Island) — lowland (in the far south) to low-alpine peat bogs. Also Tasmania.

LENTIBULARIACEAE

UPLANDS AND MOUNTAINS

UPLANDS AND MOUNTAINS

Mountains dominate the New Zealand landscape and profoundly affect climate and vegetation far beyond their immediate bounds. The dramatic moisture and vegetation gradients from the wet west to the dry east coasts are the most obvious evidence of this.

The different types of parent rocks that mountains are built of are important in their stability, the nutrients in their soils and hence the sorts of plants they support. Throughout the country, parent materials are as varied as the volcanics of the central North Island massifs, the easily fragmented greywacke of the axial ranges of both main islands, and the more stable schist or granite plateaux and monoliths of Otago and Fiordland.

Mountain climates are also far from uniform, and differences are reflected in their vegetation. Features they all share, however, are a shortened growing season and a mercurial changeability, which can see temperatures plummet and gales spring up in a matter of minutes, while frosts may occur in any month of the year.

Wind is a dominant force in mountain climates, as the dwarfed mat, cushion and rosette growth habits of many alpine plants testify. The average daily windrun on the summit of Central Otago's Old Man Range at 1500 metres altitude is around 450 kilometres — on a world scale a very high figure for such a low elevation. On this windswept schist plateau, soil is frozen for up to six months of the year and air frosts may occur in midsummer. These rigorous conditions permit the existence of only dwarfed cushion- and mat-forming plants similar to those of the climatically comparable Arctic tundra.

Some mountains present their own peculiar problems to plants: the steep, mobile stony surfaces of South Island greywacke mountains are inhabited by some 25 plant species specially adapted to conserve water and tolerate talus movement and daily temperature extremes. So specialised are these scree dwellers that they occur nowhere else. Their adaptations include such features as summer-green habit, succulent, easily shed leaves, and stout or extensive underground storage organs. The bluish-grey waxy bloom coating the leaves of many scree species perfectly merges with their shingly background as well as reducing water loss.

More than 500 species of flowering plants, or over 25 per cent of the flora, grow only in the alpine zone, that is, the thousand or so metres between tree-line and the altitude to which snow thaws in summer. (A few very hardy species in the genera *Ranunculus*, *Parahebe* and *Hebe* extend into the nival zone where snow lies permanently, surviving in snow-free crevices.)

G.R.

Dolichoglottis scorzoneroides.

G.R.

Leucogenes leontopodium.

J.F.

Euphrasia cockayneana.

More than 90 per cent of New Zealand's alpine plants are endemic — a remarkably high figure when the extreme youth (only about two million years) of our main mountain ranges is considered. The flora as a whole comprises about 80 per cent endemic species. Scientific debate continues as to how so many species could have evolved in little more than two million years, especially when prior to this mountain-building period New Zealand existed as a low-lying archipelago with a subtropical climate.

It has been suggested by Dr Peter Wardle (1978) that ancestors of today's alpine species lived in cooler uplands at the southern end of this now submerged archipelago, ready to occupy the newly formed mountain niches when these became available. He also conjectured that some of the showy-flowered plants of the subantarctic islands may be relicts from this early flora. The presence of the majority of alpine species in the South Island's higher and more extensive mountains could also be taken as support for the existence of a temperate southern land extension during the subtropical Tertiary; colonisation of the more distant North Island mountains has taken correspondingly longer and is reflected in their smaller alpine floras.

However, taxonomic studies of New Zealand's *Ranunculus* species cast doubts on these ideas. Dr F. D. F. Fisher (1965) found that our alpine ranunculi were more closely related to those of mountains in Australia and South America than to our own lowland species.

Throughout the country, alpine vegetation types vary considerably in species composition and altitudinal range, but in general the subalpine and low-alpine zones are dominated by a mixture of tussock-forming grasses, shrubs and large herbs. Important here are the grass genera *Festuca*, *Poa*

R.M.

Aciphylla scott-thomsonii.

and *Chionochloa*, herbs in *Aciphylla*, *Anisotome*, *Astelia*, *Bulbinella* and *Celmisia*, and shrubs such as *Cassinia*, *Hebe* and *Dracophyllum*.

Thus we have the plant association termed snow tussock/scrub and extending up above this in wetter regions is snow tussock/herbfield. Important tussock species are the broad-leaved snow tussock (*Chionochloa flavescens*), red tussock (*C. rubra*) and midribbed snow tussock (*C. pallens*).

In the drier eastern mountains of the central and southern South Island, large herbs and shrubs are not so abundant and the narrow-leaved snow tussock (*C. rigida*) and slim snow tussock (*C. macra*) form almost pure golden swards. Much of the eastern South Island hill country was once clothed in this snow-tussock grassland, but more than a century of burning and grazing has severely weakened the tussocks, accelerated erosion and led to the increase of unpalatable species in some areas. These include some tussock grasses such as alpine fescue (*Festuca matthewsii*), blue tussock (*Poa colensoi*) and bristle tussock (*Rytidosperma setifolia*), as well as herbs such as golden spaniard (*Aciphylla aurea*), false spaniard (*Celmisia lyallii*) and cotton daisy (*Cèlmisia spectabilis*).

In wetter regions large herbs alone may dominate some low-alpine areas, forming the garden-like herbfields that are such a magnificent sight during midsummer. The Mount Cook 'lily', *Ranunculus lyallii*, is probably the best known of our mountain herbs and often grows in association with other handsome species such as *Ranunculus insignis*, *Anisotome haastii*, *Aciphylla scott-thomsonii*, *Ourisia macrophylla* and *Gentiana bellidifolia*.

Tarns, seepages and other poorly drained sites are common in higher rainfall regions, and peat bogs in which cushion plants dominate are characteristic of the cool, moist, misty southern South Island mountains. Prominent genera of alpine wetlands include *Donatia*, *Drosera*, *Phyllachne*, *Celmisia*, *Dracophyllum*, the sedges *Carex*, *Carpha*, *Oreobolus* and *Gaimardia*, and the rush, *Juncus*.

In the high alpine zone, plant cover is much sparser, soils are meagre, and most plants grow as low rosettes, mats or cushions. The term fellfield is used for mountain slopes where stable surface rocks are interspersed with dwarf plants.

At the limits of plant growth, around the permanent snow-line, are species which can tolerate some degree of snow cover. Plants of snowbanks make

J.F.

Wahlenbergia albomarginata.

J.B.

Ourisia macrophylla.

G.R.

Celmisia incana.

the best possible use of their very short growing season; *Caltha obtusa*, for example, comes into full flower within a week of snowmelt, its buds having been fully formed the previous autumn.

Many species of our alpine plants have very localised distributions — a result of their recent evolution and inability to spread far from their highly dissected, remote ranges. Other species are notable for their disjunct distributions, with populations several hundred kilometres apart despite apparently favourable intervening habitats. This occurs most frequently between north-west Nelson and Fiordland, and the likely explanation is climatic. Central and southern Westland were so heavily glaciated during the Pleistocene that many species became extinct there, and some, such as *Celmisia traversii*, have not yet recolonised these areas from their refugia in Nelson and Fiordland.

Endemism is high (90 per cent) among alpine species but low among genera — only about 10 of the 130 alpine genera are endemic. Several, such as *Ranunculus*, *Gentiana*, *Viola*, *Poa* and *Caltha* are more or less cosmopolitan, while most are shared with other southern lands — Australia (*Craspedia*, *Pimelea*, *Herpolirion*), South America (*Acaena*, *Ourisia*, *Gunnera*) and South Africa (*Bulbinella*, *Cotula*, *Geranium*).

In keeping with the pattern shown by the flora as a whole, the flowers of most alpine species are white or yellow, although some, such as *Gentiana*, *Parahebe*, *Wahlenbergia* and *Euphrasia*, have petals veined or tinted with brighter colours. In some species of *Celmisia*, *Chionohebe* and *Ourisia*, petal undersides are coloured so that the buds are brighter than the opened flowers.

As with the rest of the flora, the absence of colourful flowers in most alpines is believed to be related to a lack of specialised pollinating insects. Flies are the main pollinators, with beetles, moths and short-tongued bees playing lesser roles. The short-tubed white or yellow flowers so common among our alpines attract a wide range of these unspecialised pollinators, and it is likely that such promiscuous pollination contributes to the prevalence of hybridism within and even between some alpine genera.

The importance of an intact vegetation cover in mountains for soil conservation in lowlands has often been overlooked in the past, and our present legacies of soil erosion and instability are a direct result. Threats to the flora and stability of our mountains continue to come from aggressive exotic weeds, grazing animals and poor land husbandry.

The following also occur in uplands and mountains: *Aporostylis bifolia, Caladenia lyallii, Cassinia vauvilliersii, Donatia novae-zelandiae, Dracophyllum traversii, Drosera spathulata, Drosera stenopetala, Hoheria lyallii, Hymenanthera alpina, Leptospermum scoparium, Lyperanthus antarcticus, Metrosideros umbellata, Nothofagus solandri* var. *cliffortioides, Parahebe catarractae, Pernettya macrostigma, Phormium cookianum, Prasophyllum colensoi, Thelymitra longifolia, Thelymitra venosa, Utricularia monanthos, Viola cunninghamii.*

R.W.

Aciphylla aurea

Aciphylla aurea

Speargrass, golden spaniard

Common in upland tussock grasslands and rocky places throughout most of the eastern South Island, golden spaniard has carried a reputation for formidability ever since Europeans with their animals first settled in these areas. Rosettes of narrow, rigid, sword-like leaves, up to 700 mm long, grow out from a central taproot, and the very sharp, spiny tips of the leaves are capable of inflicting painful cuts.

In flowering years, *Aciphylla aurea* produces spectacular spire-like flower stalks up to one and a half metres long and also armed — with sharp spines concealed amongst umbels of small yellowish flowers. The reason for such armoury remains unclear, especially in view of New Zealand's lack of native browsing animals. Some biologists have conjectured that browsing pressure from moas selected the unusual *Aciphylla* morphology. Today its only enemy is the large, nocturnal native moth *Graphania nullifera*, which lays eggs in the inflorescences. The larvae burrow down through the stout stalk and into the taproot, often killing the plant.

South Island — montane to subalpine tussock grassland, tussock-scrub and rocky places from Nelson to northern Southland.

UMBELLIFERAE

C.E.B.

Aciphylla colensoi

Aciphylla colensoi

Unlike most of New Zealand's 40 *Aciphylla* species, *A. colensoi* is not confined to the South Island. Its stiff, glaucous, spiny-tipped leaves are up to half a metre long, with prominent red or orange midribs and a pair of pungent stipules at their bases.

From the centre of the metre-tall leafy rosette there grows a stout flower stalk, two and a half metres tall and impressively armed with sharp-tipped, broad-sheathed bracts, which each subtend umbels of sweet-scented pale yellow flowers during November and December. Like most native speargrasses, plants bear flowers of one sex only; those with female flowers have, by February, a crop of pale parsnip-like seeds.

North and South Islands — moist sites in subalpine scrub, herbfield and tussock grassland from East Cape south to mid-Canterbury.

UMBELLIFERAE

D.H.

Anisotome haastii

Anisotome haastii

This robust tufted herb grows to half a metre or more tall, with long-stalked, finely divided, carrot-like leaves, each leaflet tipped by a fine hair.

Flat-topped clusters of tiny white flowers appear between October and February, each about 70 mm across and subtended by a broad, ribbed bract. As with speargrasses (*Aciphylla* species), male and female flowers are borne on separate plants; an effective though extreme way of ensuring cross-pollination. Between December and March, inflorescences on females bear 5 mm long, pale, dry fruits similar to those of speargrasses and other carrot family plants.

Highly palatable to introduced grazing animals, this attractive herb is now much less common that it once was in upland tussock grasslands and herbfields.

South Island — subalpine to low-alpine scrub, tussock grassland and herbfield along and west of the Main Divide.

UMBELLIFERAE

R.M.

Bulbinella gibbsii var. *balanifera*

Bulbinella gibbsii var. *balanifera*

This species differs from *B. hookeri* in its smaller size (about half a metre tall), its strongly arched older leaves, and in its persistent flower petals remaining pressed against the maturing fruit. Flowers appear over January and February, fruits between February and June.
North and South Islands — montane to subalpine well-drained rocky sites in tussock grassland, herbfield and scrub in the mountains of the southern North Island (Ruahine and Tararua Ranges) and in high-rainfall regions of the western South Island south of Arthur's Pass.

COMPOSITAE

B.G.

Bulbinella hookeri

Bulbinella hookeri

Maori onion, bog lily

This is one of the few native plants that have spread as a result of pastoralism — it is unpalatable to stock, and survives burning because of its tuberous roots. Its yellow flower spikes are conspicuous on wet sites in montane and low-alpine tussock grasslands during early summer. In spring its storage roots send up tufts of fleshy, strap-shaped brownish green leaves, each up to 30 mm wide and 300 mm or more long.

Plants are robust, usually being 200 to 500 mm tall, while a few individuals may reach one metre. From November, slender flower stalks up to 800 mm long emerge, their top 300 mm or so bearing a dense raceme of short-stalked, golden-yellow flowers, each up to 8 mm across. Flowers mature into dry, brown pear-shaped capsules, each about 10 mm long.

New Zealand's six endemic species of *Bulbinella* are the only members of the genus outside South Africa.
North and South Islands — bogs and wet areas in tussock grassland and herbfield of the central North Island mountains. South Island mountains from Nelson and Marlborough to North Canterbury.

LILIACEAE

J.B.

Bulbinella hookeri

Caltha obtusa

Celmisia semicordata

Caltha obtusa

White caltha

The flower buds of this small alpine herb are formed in autumn and spend the winter under snow, ready to open with the spring thaw. *Caltha obtusa* is among the first alpines to flower. Its 'double' leaves are in fact due to two basal lobes folded above the main part of the small (10 mm across) squarish blade. The five-petalled, star-like, sweetly scented flowers are each about 20 mm across. This low-growing and rather fleshy cousin of the buttercups (*Ranunculus*) forms mats up to half a metre or more across by means of stout, pale rhizomes.

South Island — low- to high-alpine snowbanks and wet sites in central and southern regions.

RANUNCULACEAE

Celmisias

New Zealand is the home of 58 of the world's *Celmisia* species, with only four in Tasmania and Australia's eastern mountains. Of the native species, all but four are mountain dwellers, growing in a number of different habitats from timberline to just below that part of the alpine zone where snow lies permanently. Size and habit vary from cushion formers such as *Celmisia sessiliflora*, comprising numerous small rosettes of short, overlapping leaves, to single rosettes or tufts such as *Celmisia semicordata*, whose leaves may be nearly half a metre long.

Although their daisy-like flower heads are very similar in design and colour, they vary in size from 10 to 100 mm across. *Celmisia mackaui*, with its pinkish-mauve ray florets, is an exception to the general pattern of a central circle of yellow disc florets and an outer ring of white ray florets. Sometimes flower heads are more or less stalkless, as in *Celmisia argentea* and *sessiliflora*, but usually they are borne on white woolly stems, which in some species are up to 400 mm long.

Hybridism between species is widespread, although most hybrids are sterile or of low fertility, so that backcrossing with parent species does not occur and species remain distinct.

Celmisia semicordata

Largest in its genus, this robust herb has sword-shaped leaves that may grow to half a metre long and 80 mm wide. During December and January, each rosette bears ten or more flower heads on stout, white, tomentose stalks up to 400 mm long. Flower heads, too, are bigger than those of any other *Celmisia* — up to 100 mm across — and contain more than 800 florets. With such attributes it is not surprisingly a popular species with alpine plant collectors here and in Britain, and in parts of Ireland will even naturalise. Three distinct subspecies have been described, which differ in leaf dimensions, hairiness and colour as well as flower characters.

South Island — subalpine scrub, snow-tussock grassland and herbfield on all but the driest mountains south of Nelson.

COMPOSITAE

P.N.J.

Celmisia sessiliflora

Celmisia sessiliflora

The grey-green cushions or mats (up to a metre wide) of this species carpet damp places in herbfield and tussock grassland throughout the South and Stewart Islands. Highly branched woody stems are completely hidden by the tight rosettes of thick, stiff, white-felted leaves at their tips. Leaf length varies with locality, being largest in the northern South Island (20 mm long), and becoming progressively shorter further south (down to 10 mm).

Flower heads (10 to 20 mm wide), nestling among the leaves, have very short, barely visible stalks (*sessiliflora* translates as 'stalkless flowers'). However, their stems elongate as fruits ripen, and the small, fluffy-floated seeds easily become airborne.
South and Stewart Islands — damp sites in tussock grassland and herbfield, as well as in bogs, snowbanks and fellfield, from the subalpine to high-alpine zones.

COMPOSITAE

E. & J.K.

Celmisia spectabilis

Celmisia spectabilis

Cotton daisy, tikuma

Like *Celmisia lyalli*, this species is an early colonist on disturbed sites in areas such as southern Canterbury, where rosettes of its variety *magnifica* occupy thousands of hectares of hill country. In fact, the abundance of *Celmisia spectabilis* can be an indication of depleted vegetation.

Rosettes may occur singly or in patches a metre or more across. Even without flowers it is a striking plant: with very thick, shining, leathery leaves (50–200 by 10–25 mm), which are yellowish-green above and covered below in a deep woolly felt of buff hairs.

This is the most widespread *Celmisia*, occurring in a range of mountain habitats through most of central New Zealand. Three distinct varieties are recognised, which differ in size and distribution.
North and South Islands — tussock grassland, herbfield and fellfield from the central North Island south to Canterbury.

COMPOSITAE

R.M.

Celmisia traversii

Celmisia traversii

Appearance and distribution are distinctive in this species: the rusty-brown hairs of the leaf undersides and margins distinguish it and *Celmisia cordatifolia* from all others in the genus. *C. traversii* has an odd discontinuous distribution — occurring in the high-rainfall regions of north-west Nelson and in Fiordland, but not along the intervening mountain chain.

Particularly handsome and sought after by collectors, *C. traversii* has large, thick leaves (150–300 by 40–60 mm) that are hairless and dark green above, but below are covered in a soft, velvety, rusty-brown felt of hairs, with a band of longer hairs around the blade edge forming a conspicuous shaggy margin. Midribs and leaf stalks are purple.

During December and January, one or a number of stout, rusty velvet flower stalks (up to 400 mm long), appear from the centre of each tuft. The narrow bracts on flower stalks are also covered in brown hairs and contrast pleasingly with the (40 to 60 mm wide) white-rayed heads.
South Island — low-alpine moist sites in tussock grassland and herbfield in north-west Nelson, southern Fiordland and western Southland.

COMPOSITAE

A.F.M.

Chionohebe ciliolata

Chionohebe ciliolata (=*Pygmea ciliolata*)

The six species of this endemic genus are all mountain dwellers and were formerly placed in *Pygmea*. Relatives of hebes, parahebes, ourisias and eyebrights, they have a similar flower structure with four or five petal lobes and one or two pairs of purple-tipped stamens.

The specific name here refers to the fringe of tiny hairs on the margins of the small (2.5–5 by 1–3.5 mm) pale green, close-set leaves. In exposed sites plants form tight cushions similar to those of *C. pulvinaris*, which is confined to the drier mountains of the eastern South Island.

There are three varieties of *C. ciliolata*, varying in cushion size and density, and the size, shape and hairiness of leaves, as well as in their distributions. During December and January the pale green cushions are studded with small (6 to 7 mm long), tubular, five-lobed white flowers.
South Island — low- to high-alpine outcrops and fellfield of the western high rainfall mountains from Nelson to Fiordland.

SCROPHULARIACEAE

R.M.

Corallospartium crassicaule

Corallospartium crassicaule

Coral broom

Although the only species in its genus, like weeping broom this two-metre-tall species differs markedly from weeping broom in habit — its stout (10 mm wide) cylindrical branches are stiffly erect.

Like most other native brooms, adults lack true leaves; presumably an adaptation that assists survival in the drought-prone eastern South Island hill country. Another xeromorphic feature is the location of stomata (minute pores that control evaporation). In coral broom these are buried deep in the hair-lined, longitudinal grooves of the branches.

During December, dense clusters of up to 20 small (7 mm long), cream, purple-veined flowers arise at notches or nodes spaced 10 to 20 mm apart along the branches. The calyx and flower stalks are densely covered in long, grey hairs, as are the 7 mm long pods, each of which has a stout beak, and contains one or two seeds. Unlike pods of most other native brooms, the valves do not part when pods are ripe, so seeds are released only when pods decay.
South Island — montane to low-alpine open, rocky sites in tussock grassland and scrub east of the Main Divide between mid-Canterbury and Central Otago.

LEGUMINOSAE

R.W.

Cotula atrata

Cotula atrata

This distinctive herb of South Island mountain screes blends so well with its rocky background that it is easily missed. In spring, underground stems send up tufts (to 100 mm long) of finely divided, grey-green, soft hairy leaves together with the remarkable 'black' (actually very dark purple) flower heads, each about 20 mm across and borne on a stout, leafy stalk up to 60 mm tall. Inner florets of the flower head have stamens, the pollen masses appearing like yellow pinheads.
South Island — screes on mountains from mid-Canterbury to north-east Otago.

COMPOSITAE

G.R.

Cotula atrata ssp. *luteola*

R.W.
Cotula pyrethrifolia var. *pyrethrifolia*

Cotula pyrethrifolia var. *pyrethrifolia*

Highly branched stems of this aromatic creeping herb send up shoots of rather fleshy (10–20 by 4–15mm) leaves, which are deeply cut into two to five pairs of lobes. Plants form loose patches one metre or more across. During December and January, slender stalks (50 to 100mm long), each arising in a leaf axil, bear a single head (8 to 16mm wide) of numerous tiny cream florets. Florets may be all male, all female or a mixture of both.
North and South Islands — subalpine to high-alpine moist open sites in snow-tussock grassland, herbfield and fellfield from the Tararua Range south to Canterbury and central Westland.

COMPOSITAE

Craspedia incana G.R.

Craspedia incana
White woollyhead

The white woolly leaves of this alpine herb die down each winter. The leaves of the basal rosettes are up to 100mm long and 30mm wide, while those on the sturdy, woolly flower stalks are shorter and narrower. Tiny yellow flowers are tightly packed into dense heads up to 30mm across, and develop into fluffy seeds, each with a hairy float that helps it to become airborne.
South Island — screes of drier mountains east of the Main Divide from central Marlborough to Otago.

COMPOSITAE

Craspedia incana G.R.

G.R.
Dolichoglottis scorzoneroides hybrid.

Dolichoglottis scorzoneroides (=*Senecio scorzoneroides*)

This slender herb is transformed during December and January with the appearance of large daisy-like flower heads. Each head is up to 60mm wide and comprises a central circle of yellow disc florets and an outer ring of elegant white petals. Leaves are long (to 250mm) and narrow (20mm wide), rather fleshy and usually slightly hairy. Plants are about 300mm tall.

Hybridism with the similar but yellow-rayed *Dolichoglottis lyallii* has given a range of intermediate forms, whose ray florets vary from cream to pale pink.
South and Stewart Islands — low- to high-alpine permanently wet sites, open sites in creek beds, around tarns, streamsides and waterfalls in tussock grassland, herbfield, scrub and fellfield throughout the wetter regions.

COMPOSITAE

A.F.M.

Dracophyllum menziesii

E. & J.K.

Epilobium glabellum

Dracophyllum menziesii

Pineapple scrub

The generic name means 'dragon-leaf', alluding to the resemblance of the leaves to *Dracaena*, the dragon tree of Tenerife.

New Zealand's 34 dracophyllums are endemic, the genus having diversified more in this country than anywhere else: only a very few species occur in Australia and New Caledonia. You are likely to see one or more species wherever you go in forest and scrub in New Zealand. Dracophyllums range from the low cushions and mat formers of alpine bogs, to upright shrubs and trees.

Although classified in the family Epacridaceae, a Southern Hemisphere offshoot of the Ericaceae, dracophyllums' parallel-veined leaves, with their broad, sheathing bases, give them a more than superficial resemblance to monocots such as cabbage tree. The erica-like, bell-shaped flowers, however, reveal their affinities to the heath family, and those of *D. menziesii* are the showiest of the genus, reminiscent of pink-flowered *Pieris* cultivars: sparingly branched panicles, about 100 mm long, bear dense clusters of creamy white flowers (each about 10 mm long), with pink, folded-back petal tips. Flowers appear during November and December and by March have matured into small, dry, brown capsules.

The resemblance of the stiffly arching leaf clusters at the branch tips to pineapple tops accounts for the common name of this species. Another common name given this and other species of the larger dracophyllums is turpentine shrubs — coined by the early settlers on account of their terpene-rich wood, which burns very well even when wet.

South Island — subalpine scrub, shrubland and herbfield of south Westland, western Otago and Fiordland.

EPACRIDACEAE

Epilobium glabellum

This is the willowherb most often seen in alpine regions, where it occurs in a variety of habitats. In favourable sites stems can reach 350 mm and are often woody at the base. Purple or reddish branches bear the glossy scalloped leaves (10–15 by 5 mm) which are also frequently red-tinged, especially below.

White or pink solitary tubular flowers appear between November and March and ripen into the slender capsules so characteristic of the genus. Seeds with hairy parachutes float free when the four capsule valves split apart from the apex.

North, South and Stewart Islands — lowland to high alpine sites in a variety of habitats (e.g. tussock grassland, riverbeds, herbfield and fellfield).

ONAGRACEAE

E. & J.K.

Epilobium glabellum

Euphrasia cuneata E. & J.K.

Forstera bidwillii E. & J.K.

Euphrasia cuneata

New Zealand eyebright

Largest of the native eyebrights, this perennial species can grow to more than half a metre tall, and develop a rather woody base, although it is usually less than 200 mm in the alpine zone. The small (5–25 by 3–10 mm), leathery leaves are arranged in pairs at nodes and are spaced well apart. One to three pairs of teeth dissect the thickened flat margins and each leaf has a triangular terminal lobe.

Flowering is prolific and occurs between January and March. Although inflorescences are often highly branched where the species grows in coastal and lowland sites, they are usually unbranched in plants in the alpine zone. Flowers are snapdragon-like, with an unequally lobed corolla and two pairs of red-brown anthers (one pair long-stalked, one short-stalked). The 15 to 20 mm long flowers have a shorter, two-lobed upper lip and a larger, three-lobed lower lip. The yellow petal bases form a conspicuous throat. The flowers mature into narrow capsules containing numerous pale brown seeds.

North and South Islands — coastal to low alpine, open rocky places and along streamsides of red tussock grassland, herbfield or fellfield from East Cape to north Canterbury.

SCROPHULARIACEAE

Euphrasia cuneata E. & J.K.

Forstera bidwillii

This is one of four New Zealand species in a small genus of herbs and softly woody shrubs found in New Zealand, Tasmania and South America. The stems of this hairless subshrub sprawl, forming roots near their bases as well as growing erect to about 150 mm tall. Glossy, leathery, oblong leaves (6–12 by 2–3 mm) are stalkless and slightly overlapping, with no distinct midrib.

Between December and April, each erect, leafy stem produces a single, slender stalk with one to three flowers at its tip. Flowers are 6 to 10 mm long and each has five to seven oval white petals, a two-lobed stigma, a pair of short stamens, and in some plants, a dark red, central eye. A variety, *densifolia*, has larger, more closely set leaves and larger flowers (to 15 mm long).

North and South Islands — montane to low-alpine grassland, herbfield, rocky places and open scrub mainly from the central North Island mountains to Nelson.

STYLIDIACEAE

J.B.

Gentiana bellidifolia

A.F.M.

Gentiana corymbifera

Gentiana
Gentians

Gentians are among the last of the alpine plants to flower, and their blooms are sometimes buried by early snowfalls in late summer and autumn. Most of the 24 native species grow in the mountains, and five occur only on outlying islands.

Apart from the colourful pink or blue flowers of some subantarctic island species, New Zealand gentians have white flowers, unlike their showy blue-flowered Northern Hemisphere counterparts. Our gentians also lack the lobes between their petals that are such a distinctive feature in their Northern Hemisphere cousins. A further difference is that nectar in New Zealand gentians is secreted by glands on the petals, not by glands at the base of the ovary.

The flowers of our native gentians are cup-shaped and deeply five-lobed, their leaves are undivided, hairless and crowded into a basal rosette, and stems stand upright only at flowering.

Gentiana bellidifolia

This attractive herb, which is widespread in North and South Island mountains, varies considerably with locality and site.

The rather fleshy dark green to reddish leaves of its basal rosette are larger (to 15 by 7 mm) than those attached to flowering stems. During March or April rosettes are overtopped by single or branched flower stems up to 150 mm tall, which bear tight, flat-topped heads of two to six stalked, bell-shaped flowers each about 20 mm across. Petals sometimes have fine purple veins.

The South Island variety, *australis*, is stouter and shorter with more prolific and larger flowers.

North and South Island — widespread in tussock grassland and herbfield in mountains from Mt Hikurangi south. Variety australis is confined to fellfield along the South Island Main Divide.

GENTIANACEAE

Gentiana corymbifera

This stout herb may grow to half a metre tall from a deep, stout (to 30 mm wide) taproot and a basal rosette of thick, rather fleshy spatula-shaped leaves (50–150 by 10–20 mm). The deep, reserve-rich taproot enables this species to persist in very depleted eroding grasslands, where plants in flower are all the more conspicuous.

A single, stout, purplish flower stalk arises from the centre of the (usually) single rosette and branches in its upper parts to form several tiers of compact, flat-topped heads. Each is about 150 mm wide and contains many 20 mm wide creamy white or mauve-tinted flowers. Flowering takes place between January and March and by late autumn the large flower heads, which are so conspicuous in South Island tussock grasslands, have matured into papery, pale brown, narrow seed capsules.

Gentiana corymbifera is usually monocarpic, i.e. plants flower just once before they die — a result of exhausting the food reserves built up over several years and of the mode of flowering, whereby the flower stalk arises from the plant's growing point.

South Island — montane to high-alpine dry tussock grassland of inland mountains from Marlborough to Otago.

GENTIANACEAE

J.F.

Gentiana patula

Gentiana patula

This species is somewhat like *G. bellidifolia* but has wider basal leaves (to 20mm) and longer flower stalks — up to half a metre tall. Sprawling stems of the basal rosettes rise at their tips to bear, during January and February, dense clusters of white or mauve-tinted flowers each up to 25mm across.
North and South Islands — wet sites in montane to low-alpine tussock grassland from the Tararua Range south, but occurring mainly east of the Divide in the South Island.

GENTIANACEAE

J.F.

Gentiana patula

G.R.

Geum uniflorum

Geum uniflorum

This herbaceous cousin of Europe's *Geum avens* has a surprisingly stout, woody base, which sends off creeping and rooting shoots, each ending in a leafy rosette. Leaves are compound with one or two pairs of small leaflets and a larger (20–40 by 15–30mm) terminal leaflet — all with toothed, hairy margins. Typical of many rose family plants, there is a pair of brown, papery stipules at the leaf bases.

During January and February, long (up to 150mm), hairy flower stalks appear, carrying a single white-petalled flower about 25mm wide. By late February or March, each has matured into a fruiting head sitting in the persistent red-tinged calyx. Styles of the separate carpels persist in fruit and colour as fruits ripen, giving the whole head a reddish hue.
South Island — widespread in damp, shady sites in mixed tussock-herbfield in the low-alpine zone of wetter mountains.

ROSACEAE

G.R.R.

Hebe macrantha var. *macrantha*

Hebe macrantha var. *macrantha*

This species has the largest flowers of any New Zealand *Hebe* — up to 25 mm wide, with the four pure white petals fashioned into a deep cup, rather like a gentian flower. Flowers are borne in groups of two to six just below the stem tips, but are so large that they often hide the tips. The leaves, too, are unmistakable — very thick, elliptic and blunt, with incised margins. They measure up to 30 mm long by 15 mm wide.

In the wild this species is a rather straggly shrub, 300 to 600 mm tall, although in cultivation it can be kept compact by pruning after flowering.

Another variety, *brachyphylla*, with smaller, more rounded leaves occurs from Nelson and Marlborough to North Canterbury.

South Island — damp, rocky sites in tussock, scrub and herbfield from Canterbury to Fiordland.

SCROPHULARIACEAE

R.M.

Hebe macrantha var. *macrantha*

D.H.

Hectorella caespitosa

Hectorella caespitosa

Firm green cushions of this dwarf shrub crouch stubbornly on the windswept tundra-like fellfields high up in South Island mountains. Cushions are up to 300 mm wide and 40 mm thick and are made up of short, stout branchlets each ending in a small (10 to 15 mm across) rosette of closely overlapping stiff, fleshy, hairless leaves (5 to 8 mm long).

During December each rosette bears up to a dozen creamy-white, five-petalled flowers which, although small (8 mm diameter), en masse almost hide the foliage. Flowers on male plants are slightly larger than on females and are easily recognised by their orange anthers. Bisexual flowers occasionally occur on plants of either sex but are borne on separate shoots. Each globose fruit nestles in a cup made by the persistent sepals but its three lobes do not gape apart when ripe, and so seeds escape only when the fruit wall rots away.

South Island — high-alpine fellfield and cushion vegetation of mountains from Arthur's Pass south along the Main Divide to Fiordland in the west and Central Otago in the east.

HECTORELLACEAE

J.B.

Helichrysum bellidioides

Helichrysum bellidioides

Everlasting daisy

The slender leafy stems of this dainty plant sprawl for half a metre or more, forming loose patches among tussocks or shrubs. Green or grey-green above, the small (4–10 by 2–8 mm), neat, hair-pointed leaves are covered beneath in a thin felt of silky white hairs.

Leafy stems, up to 100 mm long, bear prominent flower heads (up to 30 mm across), conspicuous and charming with their white, papery, persistent enclosing bracts, which make this species 'everlasting'.

North, South and Stewart Islands — lowland to low-alpine scrub, grassland and open sites from Mt Egmont and East Cape south.

COMPOSITAE

A. & S.P.

Helichrysum bellidioides

R.M.

Helichrysum coralloides

Helichrysum coralloides

Except when it is in flower, the botanical allegiances of this low shrub are difficult to discern. Its numerous stout branches are covered by small, dark green, scale leaves, which are densely lined on their inner surface with white silky hairs. The protrusion of these hairs beyond the leaf margins prominently outlines each leaf in white. The upper half of the scale leaves' outer surface is thickened and shiny.

Although it can grow to 600mm tall in favourable sites, under windy, exposed conditions this shrub is smaller and adopts a more rounded, compact, cushion-like habit.

During December or January, 6mm wide, sessile, pale yellow to cream flower heads, each comprising 20 to 40 tiny tubular florets, appear at the branchlet tips.

South Island — rock outcrops on the dry mountains of Marlborough south of Blenheim.

COMPOSITAE

A. & S.P.

Leucogenes grandiceps

Leucogenes grandiceps

South Island edelweiss

Clumps of this well-known wiry-stemmed alpine can be found sprawling over rocks throughout most South and Stewart Island mountains. Its overlapping, white, hairy leaves are up to 10mm long and 4mm wide. It is the petal-like woolly bracts surrounding the clusters of yellow flower heads that make this plant so conspicuous in the rocky sites it frequents.

South and Stewart Islands — rock outcrops on mountains from the low-alpine to the high-alpine zone.

COMPOSITAE

J.B.

Leucogenes leontopodium

Leucogenes leontopodium

North Island edelweiss

The similarity between this striking alpine herb and the European edelweiss (*Leontopodium alpinum*) is alluded to in the specific name. A more robust plant than *L. grandiceps*, it has longer, woodier stems and grows up to 150mm tall. The sessile, pointed leaves are thickly clad above and below in shining, silvery hairs, and are so densely arranged as to almost hide the stems.

Flower heads appear between November and March, and the heads themselves and the stalks bearing them are both larger than in *L. grandiceps*: flower stalks are up to 100mm tall and heads to 65mm across. Heads comprise eight to 15 central clusters of tubular yellow florets surrounded by up to 20 white, densely woolly, petal-like bracts.

Despite its common name, this edelweiss occurs in the South Island (albeit sparsely) as well as the North.

North and South Islands — low- to high-alpine rocky outcrops from Mt Hikurangi in the North Island south to Nelson and Marlborough.

COMPOSITAE

Montia australasica

Montia australasica (=*Neopaxia australasica*=*Claytonia australasica*)

This unassuming creeping herb is one of our most widespread native plants, ranging from lowland streamsides to high-alpine snowbanks and occurring throughout most of the country.

Between November and January it can be found covered with white or pale pink 20mm wide, star-shaped flowers. The narrow fleshy leaves vary greatly in length (10 to 50mm) and occur singly or in clusters.

North, South, Stewart, Auckland and Campbell Islands — lowland to high-alpine sites but most abundant in damp montane and subalpine habitats, especially grassland and herbfield. Also Tasmania and Australia.

PORTULACACEAE

Montia australasica

Myosotis pulvinaris

Myosotis pulvinaris

Soft, grey, 100mm wide cushions of this native forget-me-not hunch atop windswept rocky fellfields on high Otago tablelands.

About half our 34 native forget-me-nots are mountain dwellers and a few of these, such as *M. pulvinaris*, have a cushion habit. In *M. pulvinaris* the short (30mm) erect stems are unbranched and are covered with small silky-hairy broadly oblong leaves, each up to 7 by 5mm.

During November and December cushions are spangled with numerous dainty (6mm across), white tubular flowers borne singly at the stem tips. The corolla tube is folded inwards at its throat to form conspicuous 'scales' or 'glands', and in this species four anthers just peep out at the end of the corolla tube.

South Island — cushion vegetation of higher central and western Otago mountains with a few populations reaching northern Southland and eastern Fiordland.

BORAGINACEAE

Notothlaspi australe

Notothlaspi

This genus is one of several unique to New Zealand's mountains and its two species have a fittingly unusual appearance.

Notothlaspi australe

This close cousin of the penwiper also produces powerfully fragrant cream flower heads but differs in their being smaller and scattered over several sprawling rosettes. The slender stem branches each end in a leafy rosette and plants form loose patches up to 200mm or so across.

N. australe usually adds one rosette each year from its deep, slender rootstock. Longer-lived than the penwiper, it can flower over successive years without exhausting its reserves and dying. The flat heads of small (to 10mm wide) flowers may sometimes be recognised by scent before the plants are seen.

South Island — subalpine to high-alpine screes of mountains in Nelson and Marlborough.

CRUCIFERAE

R.W.

Notothlaspi rosulatum

Notothlaspi rosulatum

Penwiper plant

The colour and appearance of armour plating are mimicked by the tightly overlapping, fleshy, purplish-grey leaves of this crucifer family plant of South Island screes. Its circular rosettes of 20 mm long, kite-shaped, overlapping, hairy leaves were likened in Edwardian days to the penwipers made from pieces of felt stitched into neat geometrical rosettes.

During midsummer there arises from the centre of the rosettes of larger plants a sturdy flower stem up to 250 mm tall and smothered in cream, heavily honey-scented 10 to 20 mm wide flowers. Biennial like many crucifers, penwiper plants die after their once-only flowering, which usually takes place in their second summer.

The stout, fleshy taproot descends deeply below the scree surface into the moist soil below. Fine, semi-stable debris along scree margins is the most likely place to find penwipers, which are often difficult to see because of their excellent camouflage colouring and transient occurrence — plants are seldom present in large numbers in an area for more than a couple of years.

South Island — subalpine to high-alpine greywacke fine-debris screes on dry eastern mountains from Marlborough and eastern Nelson to North Otago.

CRUCIFERAE

G.R.

Ourisia macrophylla

Ourisia macrophylla

Ourisia is a genus of southern lands with 14 species in Tasmania and South America. Eight of our ten native ourisias are alpines.

This robust shade-loving herb bears a stout, conspicuous flower stalk up to 600 mm tall, which towers over tufts of large (to 150 mm long), toothed, oval leaves sent up by thick creeping and rooting stems. The five corolla lobes span 20 mm from tip to tip, and are white on their upper portions, yellow towards their bases. Each flower is borne on a slender stalk up to 50 mm long, together with three to 15 or so others in a whorl, i.e. all the stalks arise at the same level. Each inflorescence has up to eight whorls of flowers.

Four alpine varieties of *Ourisia macrophylla* are recognised; these occur from Mt Egmont (var. *macrophylla*) and the central and southern North Island mountains (vars. *robusta* and *drucei*), throughout all but the driest South Island mountains and on Stewart Island (var. *lactea*).

North, South and Stewart Islands — montane to low-alpine shady sites in tussock grassland and herbfield from Mt Egmont and the central North Island mountains south.

SCROPHULARIACEAE

G.R.

Ourisia macrophylla

D.H.

Ourisia sessilifolia var. *sessilifolia*

Ourisia sessilifolia

Although a much smaller plant than *O. macrophylla*, this species has similar flowers, which are nearly as big (15 mm across) and are borne in pairs at the end of 50 to 150 mm long, hairy stalks.

The three varieties of this species together cover most of the higher-rainfall mountains of South and Stewart Islands. They vary mainly in the size, shape and arrangement of leaves and in the structure of the coarse hairs covering them. Habitat preferences differ too, with variety *splendida* extending to the highest altitude in fellfield and snowbanks, while varieties *sessilifolia* and *simpsonii* occur in rocky open sites mainly in snow-tussock herbfield.
South and Stewart Islands — low- to high-alpine snow-tussock herbfield, snowbanks and fellfield on wet mountains from north-west Nelson south.

SCROPHULARIACEAE

Parahebe planopetiolata A.F.M.

Parahebe planopetiolata

The shining, dark green, spoon-shaped leaves (up to 10 by 6 mm) of this dainty herb are sent up by slender stems, which sprawl over damp rocky sites in a wide range of altitudes and vegetation types.

Corollas of the small flowers (each about 5 mm across) may be white, pink or lavender, and a pair of large purple anthers are clearly visible at the mouth. Flowers are borne singly or in pairs on very short stalks near the branch tips during midsummer.
South Island — damp, rocky low- to high-alpine sites in the mountains of Fiordland and west Otago.

SCROPHULARIACEAE

B.G.

Pittosporum dallii

Pittosporum dallii

Our only native pittosporum with white flowers and serrate leaves, this species is one of many worthy ornamentals in its genus.

Pittosporum dallii grows into a small, grey-barked, round-crowned tree of six metres or so. The thick, dark green leaves (50–100 by 25–35 mm) are elliptic in shape and are usually clustered toward branchlet ends.

From November to January, branch tips bear flat-topped clusters of strongly scented white flowers, each up to 30 mm across. By autumn these have ripened into woody, two-valved capsules, about 10 mm long, containing black seeds embedded in a sticky pulp and enclosed in a papery casing. In cultivation, seeds may take over a year to germinate.

Confined to a few sites in mountain forests of north-west Nelson, this *Pittosporum* is considered in possible danger of extinction in the wild, although a sanctuary for it has now been set up at Snows River within the North-West Nelson State Forest Park.
South Island — montane and subalpine forest margins and rocky sites in north-west Nelson.

PITTOSPORACEAE

D.H.

Ranunculus buchananii

G.R.

Ranunculus haastii

Ranunculus
Buttercups

About half of the 35 New Zealand buttercups are alpines and they include the most spectacular of our mountain flowers, *Ranunculus lyallii*, popularly but misleadingly called the Mount Cook lily. It is the best known; eye-catching, large, circular leaves and showy white flowers.

Although our alpine and other buttercups vary widely in their preferred habitats, size and foliage, their flowers are immediately recognisable by a similarity of design: a central cone of female carpels surrounded by spirals of the conspicuous, large stamens and yellow or white petals. This type of flower structure — where the flower parts are all unfused and spirally arranged, and the carpels sit above the other parts — is thought to resemble that of the first flowering plants, which evolved around 100 million years ago. It is interesting that the closest relatives of New Zealand's alpine *Ranunculus* species are not our lowland species, but other alpine ranunculi in Australia and South America.

Ranunculus buchananii

A true aristocrat among our buttercups, *R. buchananii* is the only species to share with *R. lyallii* the distinction of white flowers. These appear in December or January, borne on 300 mm long stems in groups of one to three, each satiny-petalled bloom up to 60 mm across. Pale green at first, the fruiting heads ripen dry and brown. Rosettes of blue-green, often silky, hairy leaves with their three-lobed, variably dissected blades, are borne on long (to 150 mm) sturdy stalks.

One of the highest-growing alpines (it reaches 2400 metres), *R. buchananii* is a summer-green herb that grows near the permanent snow-line and overwinters by means of its stout rootstock.

South Island — high-alpine fellfield and stable rock debris, now mainly restricted to inaccessible sites in western mountains south of Arthur's Pass, but reaching as far east as the Remarkables and Hector Mountains in the Central Otago lakes district.

RANUNCULACEAE

Ranunculus haastii

Well camouflaged amid their rocky background, the thick, waxy, glaucous, mauve-margined leaves of *R. haastii* arise directly from the rootstock — a stout rhizome that reaches down through the mobile shingly surface to the moist, stable, sandy soil below. Leaf blades are deeply divided, measure up to 100 mm across, and are borne on stalks 50 to 150 mm long.

Flowers give away the plants' presence when they appear — between November and January. Stems up to 150 mm long bear large (20 to 40 mm wide) flowers, with the classic glossy yellow petals of buttercups. A subspecies, *piliferus*, confined to northern Southland's Eyre Mountains, differs in having larger (to 50 mm wide) flowers and less dissected leaves, which are white and hairy below.

South Island — low- to high-alpine screes of dry greywacke mountains in the central and eastern regions from Marlborough to western Southland.

RANUNCULACEAE

J.B.

Ranunculus insignis

Ranunculus insignis

This robust (to 900 mm tall) herb, which favours shady places in the lower alpine zone, is a striking sight between November and February when it sends up 750 mm tall flower stems bearing as many as 50 large (20 to 50 mm wide), yellow-petalled flowers. It is a highly variable species: North Island populations have much larger, heart-shaped leaves (up to 160 mm wide) and many more flowers than plants at the southern end of its range in South Canterbury. The leaves of all forms, however, are unlobed, and their coarsely toothed margins are conspicuously fringed with brown hairs.

North and South Islands — subalpine to low-alpine shady sites in scrub, snow-tussock grassland and herbfield on the eastern, central and southern North Island mountains and eastern South Island mountains from Nelson to South Canterbury.

RANUNCULACEAE

J.F.

Ranunculus lyallii

Ranunculus lyallii

Great mountain buttercup, Mount Cook lily

The world's largest buttercup, this majestic summer-green herb can grow to one and a half metres tall with circular, leathery leaves up to 400 mm across and white flowers 80 mm wide. Widespread through the wetter South Island mountains, it is probably our best-known alpine flower. Its popular name, 'Mount Cook lily', is misleading, for the plant is unrelated to lilies.

Ranunculus lyallii has suffered badly from grazing by introduced game animals and the largest plants are now relatively rare — most specimens are about one metre tall with leaves 130 to 200 mm across.

Seedlings have kidney-shaped leaves, but these mature into the distinctive circular blades with their finely crenate red margins. Centrally attached stout leaf stalks are 200 to 300 mm long, and their broadly sheathing bases are covered in silky hairs. Leaves are otherwise hairless. During December and January, tall (to one metre) branched flower stems bear showy panicles of large (50 to 80 mm wide), white-petalled flowers. Each flower has 10 to 25 oblong overlapping petals surrounding golden stamens and a central green cone of spirally arranged carpels. Fruiting heads (30 mm long) of pale, hairy achenes are conspicuous by midsummer.

South and Stewart Islands — subalpine to low-alpine herbfield of the wetter South Island mountains and on Stewart Island's Mt Anglem.

RANUNCULACEAE

D.H.

Ranunculus lyallii

Ranunculus sericophyllus

Raoulia mammillaris

Ranunculus sericophyllus

In flower this perennial herb is one of the most conspicuous plants of the high-alpine zone. Between November and February, shiny, yellow-petalled, stout-stalked flowers sit just above the rosettes of very finely dissected, three-lobed leaves. Usually flowers are borne singly, but occasionally pairs are seen. Both the flowers, which may be up to 40 mm wide, and the globose, centimetre-wide seed heads, appear disproportionately large for the small rosettes. The slender (about 2 mm wide) tapered leaf segments are fringed with silky hairs.
South Island — high-alpine snowbanks and sheltered fellfields on higher mountains from Lewis Pass south.

RANUNCULACEAE

Raoulia

New Zealand is the headquarters of this genus whose name commemorates Etienne Raoul (1815–52), a botanist and surgeon with a French expedition that visited this country in 1840–41.

Raoulia comprises mat- and cushion-forming herbaceous and shrubby daisies with short overlapping leaves. Rayless flower heads without stalks nestle among the leaves. The cushion-forming species are popularly known as vegetable sheep because of their somewhat sheep-like colour and appearance from a distance. Sixteen of our 20 species reach the alpine zone. Tough and hardy, all are well adapted to life in exposed, dry, stony or rocky places, whether these be gravelly riverbeds, shattered rock fellfields or rock outcrops in tussock grassland, where there are large differences between day and night temperatures, especially during summer.

Raoulia mammillaris

Vegetable sheep

Hard, silvery oval hammocks of this remarkable daisy squat among the shattered rock of windswept slopes and ridges in stable South Island fellfields. Unlike the more widespread soft-textured *R. eximia*, *R. mammillaris* is harsh to the touch, owing to short hairs at the leaf tips. It is also smaller than *R. eximia*, with cushions rarely more than half a metre across. The tightly overlapping 'woolly' leaves of this vegetable sheep are 3 to 4 mm long and 1 mm wide. Flower heads (5 to 6 mm wide) are made conspicuous by an encircling ring of white-tipped scales.

Like other vegetable sheep in the genera *Raoulia* and *Haastia*, *R. mammillaris* is a highly branched shrub whose adjacent leafy branch tips are so tightly pressed together that they form a regular and distinctive pattern on the outer surface of the cushion. This mode of growth helps conserve water in the very dry places where vegetable sheep live; fallen leaves within the cushion act like a sponge, readily absorbing and retaining water.
South Island — high-alpine zone of dry greywacke mountains in north and mid-Canterbury.

COMPOSITAE

J.F.

Raoulia subsericea

Raoulia subsericea

This is among the commonest mat plants of South Island tussock grasslands, especially in eroding, overgrazed areas. Its loose olive-green mats are 200 mm or more across and the narrow (3–6 by 1 mm) pale green leaves are thinly covered below in a silver or gold hairy felt. Flower heads (7 to 10 mm wide) are each defined by a ring of papery, pale, petal-like, white-tipped scales.
South Island — subalpine to low-alpine tussock grassland in all but the wettest mountains.

G.R.

Stellaria roughii

Stellaria roughii

Probably the commonest of the 20 or so species specialised to live in the mobile shingle of South Island screes, *Stellaria roughii* forms loose bluish-grey tufts about 100 mm tall and up to 200 mm across.

In spring the hair-like underground stems send up shoots that branch repeatedly and are clothed in succulent narrow leaves each 20 by 4 mm. By midsummer, most branch tips sport a single 20 mm wide flower, whose leaf-like sepals partly conceal the shorter bi-lobed white petals. The sepals persist and also tend to hide the dry capsular fruits.
South Island — low- to high-alpine screes on mountains from Nelson south to western Southland.

G.B.

Wahlenbergia albomarginata

Wahlenbergia albomarginata
Harebell

In flower, harebells are some of the most noticeable small herbs of tussock grasslands. Between November and February each rosette (10 to 20 mm across) sends up a single, very slender stem, which bears just one white or pale blue, bell-shaped flower 10 to 30 mm long. Petal backs are often marked by a darker blue line.

The spoon-shaped leaves vary considerably in size, depending on the growing conditions — in dry areas they may be as short as 5 mm or, in moist sites, as long as 40 mm. Their scalloped edges are often thickened, and northern South Island populations have distinctly glaucous leaves. Rosettes of this harebell occur in patches derived from a single, spreading, wiry rhizome.

Wahlenbergia is our only native genus in the campanula family. There are nine other native species, including two other alpines.
South and Stewart Islands — widespread in lowland to low-alpine tussock grassland and herbfield.

COMPOSITAE

CARYOPHYLLACEAE

CAMPANULACEAE

GLOSSARY

achene: a dry, indehiscent, one-seeded fruit as found in e.g. clematis and most composites.

acute: sharply pointed.

adventitious: arising from an unusual position, e.g. roots from stems or leaves.

alternate: arranged singly, not in opposite pairs.

androecium: the male reproductive organs of a flower: the stamens. Cf. gynoecium.

angiosperm: the class of seed plants whose seeds are enclosed in an ovary: a flowering plant.

annual: completing its life-cycle in one year.

anther: the pollen-bearing part of a stamen, usually borne at the top of a filament.

apetalous: without petals.

apex: the tip or free end of an organ: hence apical.

aril: an appendage, often fleshy, on the outside of a seed.

awn: a stiff, bristle-like projection, especially on the glumes of grasses.

axil: the upper angle between two dissimilar parts.

bast fibres: the fibres of the phloem or inner bark.

berry: a fleshy fruit, containing a number of seeds but not a 'stone'.

biennial: living for two years and flowering only in the second.

blade: the broad, usually flat part of a leaf.

bract: a modified, often much reduced leaf, especially on a flower stem.

bryophyte: a moss or liverwort.

calyx: the outer, usually green series of floral parts consisting of separate or joined sepals.

capsule: a dry fruit, formed from two or more fused carpels, which usually splits to release its seeds.

carpel: one unit of the female part of a flower, consisting of a basal ovule-bearing ovary which is usually joined to a stigma by a stalk-like style.

column: the combined stamens, style and stigmas of orchids.

composite: a member of the daisy family, Compositae.

compound leaf: a leaf comprising two or more leaflets.

corolla: the usually showy inner series of floral parts, comprising separate or joined petals.

corymb: a more or less flat-topped inflorescence, where the long-stalked outer flowers open first.

crown: the area of the junction between root and stem.

cultivar: a variety of garden origin or a selected form from the wild maintained in cultivation.

cyme: an inflorescence in which each terminal growing point produces a flower, so that the oldest flowers are at the apex — or centre, if flat.

dicot: a dicotyledon — a plant belonging to one of the two subclasses of flowering plants. Dicot embryos have two cotyledons (seed leaves), and their true leaves are net-veined. Cf. monocot.

disc florets: the small tubular florets in the centre of the flowerheads of many composites.

divaricate: a term for a divaricating shrub, i.e. one whose branches have wide angles between them, small leaves and are intertangled.

drupe: a fleshy indehiscent fruit whose seed is enclosed in a 'stone'.

elliptic: widest at the middle and tapering to each end.

endemic: native only to a particular country or region.

entire: (of leaves) with a straight or smoothly curving margin: untoothed.

epiphyte: a plant which perches on another but is not organically connected to it.

family: a category in plant classification above the rank of genus: a group of plants sharing distinct characteristics and comprising one or more genera.

fascicled: arranged in clusters.

filament: the stalk of an anther.

flaccid: limp.

floret: a small flower, usually one of a cluster, as in the head of a daisy or the spikelet of a grass.

fruit: a ripened ovary containing one or more seeds — see **achene, berry, capsule, drupe, nut.**

glabrous: without hairs.

glaucous: of a bluish-green colour.

globose: spherical, rounded.

glume: a small chaffy or membranous bract, especially at the base of a grass spikelet.

gymnosperm: the class of seed plants whose seeds are not enclosed in an ovary e.g. podocarps, pines, cypresses and other conifers.

gynoecium: all the female reproductive organs of a flower, comprising one or more free or fused carpels.

habit: the general appearance of a plant.

head (capitulum)**:** a dense inflorescence of a large number of usually stalkless flowers as in most composites.

hood: the modified upper sepal of an orchid flower.

indehiscent: fruits not opening to release seeds, except by decay.

inflorescence: the flowering part of a plant, or the arrangement of flowers — see **corymb, cyme, head, panicle, raceme, spike, umbel.**

involucre: one or more whorls of bracts surrounding a flower head.

keel: an obvious central ridge.

labellum: the lip or middle petal of an orchid flower, which usually differs from the two lateral petals in size, shape or ornamentation.

lanceolate: lance-shaped: four to six times longer than broad, with curved sides and pointed ends, and attached at the broader end.

lax: loosely arranged or widely spaced.

leaflet: one leaf-like segment of a compound leaf.

lemma: the lower and larger member of the pair of bracts surrounding a grass flower.

linear: very narrow with parallel margins.

midrib: the central or largest vein of a leaf.

monocarpic: flowering and fruiting only once, then dying.

monocot: a monocotyledon — one of the two subclasses of flowering plants. Monocot embryos have one cotyledon (seed leaf) and their true leaves usually have parallel veins. Cf. dicot.

montane: the altitudinal zone lying above the lowlands and below the subalpine regions.

nival: growing in snow.

node: a place on a stem where one or more leaves is attached.

nut: a dry, single-seeded, indehiscent fruit with a hard, woody wall.

oblanceolate: lanceolate, but broadest above the middle and attached at the narrow end.

obovate: broadest above the middle and attached at the narrow end.

obtuse: blunt.

offset: a short, thickened, leafy stem, usually of rosette-like appearance, which develops as a lateral shoot from the base of the main stem.

oval: broadly elliptic.

ovate: the shape of a lengthwise section of a hen's egg and attached by the broad end.

ovary: the hollow basal region of the gynoecium which contains the ovules. After fertilisation the ovary develops into a fruit and the ovules into seeds.

palea: the smaller and upper member of the pair of bracts enclosing a grass flower.

palmate leaf: (e.g. pate) a compound leaf with three or more leaflets arising from the same point.

pan: a soil layer that is distinctively compact, and rather impermeable, so as to impede the passage of water and roots.

panicle: a branched inflorescence of stalked flowers.

parasite: a plant that obtains its food from another living plant to which it is attached.

pedicel: the stalk of a single flower in a compound inflorescence.

peduncle: the stalk of a solitary flower or the main stalk of a compound inflorescence.

perennate: to persist from season to season.

perianth: the calyx and corolla together.

petal: a separate member of the corolla.

petiole: the stalk of a leaf.

phyllary: an individual bract of an involucre, as occurring in the flower heads of composites.

pinnate leaf: (e.g. kowhai) a compound leaf with leaflets arranged along both sides of the stalk (once-pinnate). A twice-pinnate leaf, e.g. some *Aciphylla* spp., has the first-order leaflets themselves pinnate. A 3-pinnate leaf, e.g. some *Aciphylla* and *Anisotome* spp., has the second-order leaflets pinnate.

pistil: the female reproductive organs of a flower, consisting of one or more carpels, each comprising ovary, style and stigma. The gynoecium as a whole.

pollination: the transfer of pollen from stamen to stigma.

pubescent: clad in short, soft hairs.

pungent: ending in a stiff, sharp point.

raceme: an inflorescence of stalked flowers arising from a single axis, with the oldest flowers at the base.

ray florets: the outer florets in the flower heads of many composites. One of the petals in each ray floret is extended into a strap-like ray.

rhizome: a perennial, underground, usually horizontal stem which bears roots and leafy shoots.

rhomboid: (of leaves) more or less diamond-shaped.

rootstock: a short, erect underground stem.

rosette: a group of leaves arising close together from a short stem and forming a radiating cluster on or near the ground.

rostellum: a sterile structure separating anther and stigma in orchids.

runner: a trailing stem which roots from the nodes.

scabrid: rough to the touch because of minute, stiff projections.

scale: a small, sometimes thin, leaf, bract or flap.

sepal: a separate member of the calyx.

serrate: sharply toothed with the teeth pointing forward.

sessile: without a stalk.

sheath: the base of a leaf or leaf stalk which encloses a stem.

simple leaf: one not divided into separate leaflets.

spathulate: spoon-shaped.

spike: a raceme of unstalked flowers.

spikelet: a small cluster of one or more florets in grasses, sedges and rushes.

stamen: the male or pollen-bearing organ of a flower, consisting of an anther and its stalk.

stomata: pores in a leaf through which gases pass.

stigma: the part of a carpel which receives pollen, usually at or near the tip of the style.

stipule: a scale-like or leaf-like appendage at the base of a petiole, and usually occurring in pairs.

style: the usually elongated part of a carpel between the ovary and stigma.

subtend: to stand below but usually close to another organ, as a bract to its flower.

summer-green: dying back to underground parts in winter.

taxon (plural taxa): a category of any rank in a classification system, e.g. family, genus, species.

tepal: a perianth segment that is not clearly distinguishable as a sepal or a petal.

terete: circular in cross-section.

tomentum: a dense covering of soft short hairs; hence **tomentose**.

trifoliate: having three leaflets.

tuber: an underground perennial stem or root which is swollen with food reserves.

turgid: firm from distension with water.

umbel: a more or less umbrella-shaped inflorescence where the stalks of the individual flowers radiate from a common centre.

valve: a distinct division of a capsule.

variety: a distinct variant of a species, often with its own geographical distribution in the wild.

vascular plants: those with specialised conducting tissue. Includes ferns, fern allies, gymnosperms and angiosperms.

whipcord: a shrub in which the leaves are reduced to scales that are close-set and pressed against the stem, as in some alpine hebes.

whorl: an arrangement where organs (e.g. sepals, petals, stamens, leaves) arise at the same level around an axis.

xeromorphic: having structures (e.g. small leaves, succulence, dense hairiness, thick cuticles) which help conserve water and withstand drying conditions.

REFERENCES AND FURTHER READING

Flower Biology

GODLEY, E. J., 1979: 'Flower biology in New Zealand', *N.Z. Journal of Botany 17*: 441–466.

HOLM, E., 1979: *The Biology of Flowers*, Penguin Books, Harmondsworth, England.

LLOYD, D. G., 1985: Progress in understanding the natural history of New Zealand plants, *N.Z. Journal of Botany 23*: 707–722.

MEEUSE, B., MORRIS, S., 1984: *The Sex Life of Flowers*, Faber and Faber, London.

MOORE, L. B., IRWIN, J. B., 1978: *The Oxford Book of New Zealand Plants*, Oxford University Press, Wellington.

PROCTOR, J., PROCTOR, S., 1978: *Nature's Use of Colour in Plants and their Flowers*, Cassell, Australia.

THOMAS, B., 1981: *The Evolution of Plants and Flowers*, St Martin's Press, New York.

Conservation and Ecology

COCKAYNE, L., 1928: *The Vegetation of New Zealand*, W. Engelmann, Leipzig.

COCKAYNE, L., 1967: *New Zealand Plants and their Story* (4th edn., ed. E. J. Godley), Government Printer, Wellington.

ENTING, B., MOLLOY, L., 1982: *The Ancient Islands*, Port Nicholson Press, Wellington.

FISHER, F. D. F., 1965: 'The alpine Ranunculi of New Zealand', *N.Z. Department of Scientific and Industrial Research, Bulletin 165.*

GIVEN, D. G., 1981: *Rare and Endangered Plants of New Zealand*, Reed, Wellington.

GIVEN, D. G., WILLIAMS, P. A., 1984: *Conservation of Chatham Island Flora and Vegetation*, Botany Division, Department of Scientific and Industrial Research, Christchurch.

GODLEY, E. J., 1975: 'Flora and Vegetation', in G. Kuschel (ed.), *Biogeography and Ecology in New Zealand*, 177–229.

MARK, A. F., 1977: *Vegetation of Mount Aspiring National Park, New Zealand*, National Parks Authority, Wellington.

MOLLOY, B. P. J., SIMPSON, M. J. F., 1980: 'Taxonomy and ecology of *Pachystegia* (Compositae): a progress report', *N.Z. Journal of Ecology 3*: 1–3.

MORTON, J., OGDEN, J., HUGHES, T., 1984: *To Save A Forest — Whirinaki*, Bateman, Auckland.

NEW ZEALAND'S NATURE HERITAGE, 1974-76: Hamlyns, Hong Kong, 7 vols. This encyclopaedia includes numerous articles on the biology of native plants.

OGLE, C. C., WILSON, P. R., 1985: Where have all the mistletoes gone?, *Forest and Bird 16*: 10–13.

WARDLE, P., 1978: Origin of the New Zealand mountain flora, with special reference to trans-Tasman relationships, *N.Z. Journal of Botany 16*: 535–550.

WARDLE, P., 1979: *Plants and Landscape in Westland National Park*, National Parks Authority, Wellington.

WILSON, H. D., 1976: *Vegetation of Mount Cook National Park, New Zealand*, National Parks Authority, Wellington.

Identification

ALLAN, H. H., 1961: *Flora of New Zealand*, Volume I. Government Printer, Wellington.

COOPER, D., 1981: *A Field Guide to New Zealand Native Orchids*, Price Milburn, Wellington.

EAGLE, A., 1975: *Eagle's Trees and Shrubs of New Zealand*, Collins, Auckland.

EAGLE, A., 1982: *Eagle's Trees and Shrubs of New Zealand*, second series, Collins, Auckland.

JOHNS, J., MOLLOY, B., 1983: *Native Orchids of New Zealand*, Reed, Wellington.

MARK, A. F., ADAMS, N. M., 1986: *New Zealand Alpine Plants* (2nd revision), Reed Methuen, Auckland.
Also includes sections on alpine plant ecology.

MOORE, L. B., EDGAR, E., 1970: *Flora of New Zealand*, Volume II, Government Printer, Wellington.

POOLE, A. L., ADAMS, N. M., 1980: *Trees and Shrubs of New Zealand*, (4th impression), Government Printer, Wellington.

SALMON, J. T., 1980: *The Native Trees of New Zealand*, Reed, Wellington.
Also includes sections on forest ecology.

SALMON, J. T., 1985: *Collins Guide to the Alpine Plants of New Zealand*, Collins, Auckland.

WILSON, H. D., 1978: *Field Guide: Wild Plants of Mount Cook National Park*, Field Guide Publications, Christchurch.
This and the volume below are very useful guides to much of the flora.

WILSON, H. D., 1982: *Field Guide: Stewart Island Plants*, Field Guide Publications, Christchurch.

Cultivation and Uses

BROOKER, S. G., CAMBIE, R. C., COOPER, R. C., 1981: *New Zealand Medicinal Plants*, Heinemann, Auckland.

CROWE, A., 1981: *A Field Guide to the Native Edible Plants of New Zealand*, Collins, Auckland.

FISHER, M. E., SATCHELL, E., WATKINS, J. M., 1970: *Gardening with New Zealand Plants, Shrubs and Trees*, Collins, Auckland.

METCALF, L. J., 1987: *The Cultivation of New Zealand Trees and Shrubs* (fully revised edn.), Reed Methuen, Auckland.

INDEX

Bold numbers indicate location of subject entry. Roman numbers indicate other references including photographs.

Michael Ayrton

An Illustrated Commentary

Peter Cannon-Brookes

1978 Birmingham Museums and Art Gallery

The Publications Unit
City Museums and Art Gallery
Chamberlain Square
Birmingham B3 3DH
ISBN 0 903504 14 6

Printed by B.H. Blackwell (Printing) Ltd., Oxford

Author's Preface

Publication of this illustrated commentary on the life and work of Michael Ayrton was originally intended to coincide with the major commemorative exhibition which was shown in Birmingham in January-February 1977. It has been delayed, but in the meantime Michael Ayrton's *Archilochos Suite* of etchings has been published, and critical interest in his work continues to grow. Nevertheless, it is still extremely difficult to achieve sufficient detachment to write a commentary, let alone a critical study, when Michael Ayrton's voice still rings in our ears. Much of the account of his sculpture and the related drawings was drafted in close consultation with him, both at Bradfields and in London, and those extended discussions will always be treasured by me. I regret, however, more than I can ever express, never having discussed in depth with him his activity as a painter. Michael Ayrton's contribution to post-war British painting is beginning to be recognized, and I am only too conscious of the need for much more research before it is too late. My debt to Elisabeth Ayrton and her family is enormous, and I am most grateful to her and Prue Hopkins for their inexhaustible patience. For my account of the earliest years of Michael Ayrton's career I have also relied heavily upon the advice so generously given by Joan Foa, Michael Middleton and John Arlott, though any errors of fact or interpretation are entirely mine, and it is impossible to acknowledge fully my indebtedness to the dozens of old friends of Michael Ayrton who have discussed specific problems with me during the last three years, not least Michael Le Marchant of the Bruton Gallery, Bruton, Somerset.

Many of Michael Ayrton's works cannot at present be traced, and I have been compelled to rely upon old photographs. A certain variability in the quality of the illustrations is thus inevitable and, even for the sculpture, it has not been possible to rephotograph everything. With few exceptions, the bronzes have been issued in editions and thus the locations of only those casts in museums and other public collections are noted. However, I would be most grateful to learn of the location of any of the works described and illustrated as 'present whereabouts unknown'.

The design and production of this commentary has been undertaken by Trevor Jones and the Publications Unit of the City Museums and Art Gallery of Birmingham, and to him, to Dennis Farr, Director, and all my colleagues in the Department of Art who have supported me in both the organisation of the exhibition and the writing of this study, I extend my grateful thanks.

Peter Cannon-Brookes
City Museums and Art Gallery, Birmingham
November 1977

In 1973 David Piper made the shrewd observation that 'Michael Ayrton is in urgent need of close scholarly attention', and notwithstanding the many excellent introductions to successive exhibition catalogues, his own account of his sculptures and the generous obituaries published in 1975, that need remains urgent. Professor Jacob Nyenhuis's definitive study of the Daedalus theme, to be published shortly, will do much to clarify a major preoccupation of Michael Ayrton since 1956, but it is to be hoped that the 1977 Commemorative Exhibition will encourage the preparation of a full-scale study. In the meantime, there is an evident need for a more modest account of Michael Ayrton's creative activity, bringing together his paintings and theatrical designs as well as his sculptures and prints, and relating him to the post-war development of British art. Nevertheless, the writer on the art of Michael Ayrton is presented with a well-nigh insuperable problem in that the creator of his intended subject matter was considerably more articulate than the majority of art critics, and by definition much better informed. It is still impossible to separate the artistic creations of Michael Ayrton from the overwhelming presence of the man and his multifarious activities, and it was this very versatility which discouraged critics and art historians from venturing to prepare an extended study of his work before he was safely dead. The fluency with which he expressed his ideas in a variety of media, and his richly ironical humour, provided little encouragement for traditional forms of in-depth analysis, and it was always a brave critic who took upon himself the unremunerative task of 'explaining' Michael Ayrton's art during his lifetime. It it thus no coincidence that many of the best studies of his work, mostly in the form of exhibition catalogue introductions, were written by those not normally thought of as either art critics or art historians, and the pungent note of self-mockery in his *Autobiography of Lameich Trojan* (1972) was warning enough.

Michael Ayrton attracted a body of admirers distinguished more by their independence of thought than by any other characteristic, and in a sense this contributed to his own isolation. He neither sought nor received general critical acclaim from the mid-1940s, and Wyndham Lewis's words in June 1949, 'Michael Ayrton is one of the two or three young artists destined to shape the future of British Art', might reveal Wyndham Lewis's sensitive eye for originality and quality, but it also reveals a monumental misunderstanding of the personality of the artist. Michael Ayrton had already demonstrated an almost maverick independence which minimized the possibility that either he or his work could ever act as the focus of a specific new movement in English art. He never suffered fools or sycophants gladly, though as he was a brilliant conversationalist they rarely felt it, and his acute sense of the ridiculous often defused a potentially difficult situation. This should not be misinterpreted as intellectual arrogance but it highlights the uncomfortable

Self Portrait, 1946

1 The Vase of Flowers, 1940

position of the articulate and literate artist today. One of the legacies of the Romantic conception of the artist has been to deny him the right to be articulate, thus greatly simplifying the *modus operandi* of the critic of contemporary art. The idea of the creative artist not only being able to express in words what he is trying to achieve, but to do so fluently, is highly inimical to the pride and preconceptions of many art critics. In his *Act of Drawing* (1957) Michael Ayrton wrote that 'The artist in Western society is encouraged to distrust his intellect' and throughout his entire career his ceaselessly enquiring mind totally rejected any such restraints. In this respect his wide interdisciplinary interests and remarkable powers of analysis revealed a versatility more usually associated with the Renaissance, anachronistic in this age of specialization, and like the *uomo universale* of the Renaissance he was also a superlative craftsman. Furthermore, in these terms his colossal self-confidence was not merely understandable but a positive asset, and the constant urge to explore new means of expression a natural corollary. It is thus not surprising that for Michael Ayrton there was no primary means of expression (although drawing always remained central to his art), but rather that each creative activity cross-fertilized every other. Art critic, art historian, novelist, broadcaster, theatrical designer and film director as well as painter, sculptor and etcher, his talents were such that his activities in any single one of these fields would have assured him of an international reputation.

The pendulum has begun to swing back from the extremes of abstract art, and consequently certain fundamental problems of contemporary society begin to take on a new urgency. As visual and audio impressions received from the media replace the written word, contemporary Western society suffers from the growing illiteracy which encourages appreciation of the work of art for its formal and colouristic qualities only. Hence, in the wake of Roger Fry's demand for 'form without content', the great popularity of French Impressionism and Abstract Expressionism (since neither demands any degree of intellectual sophistication on the part of the observer). On the other hand, the intellectual content of Medieval and Renaissance art isolates them increasingly from a public which is unwilling to learn the language of Medieval imagery or that of the Classical myths and Renaissance Humanism. From 1956 one of the driving forces permeating Michael Ayrton's creative activity was the Greek myths and without a certain knowledge of Classical mythology much of the meaning of his sculptures is lost on the observer. The intense sensuality of the bronzes represents only part of what they have to offer: the rest is cerebral. Michael Ayrton himself was rarely lost for words to express precisely what, in the medium of sculpture, he was trying to achieve, and the principal rôle of the critic in the present circumstances must be to provide a running commentary.

The Ayrton family has never been distinguished for conformity or lack of imagination, and Michael has written with pride about the eccentricities of his grandfather, Professor W. A. Ayrton, F.R.S., and his experiences as Professor of Physics in the University of Tokyo. Shortly after the death of his first wife and his return from Japan in 1884, W. A. Ayrton married Hertha Marks, who shared his passion for physics and electrical engineering. A scientist of great distinction in her own right, and a lifelong friend of Marie Curie, Hertha Ayrton worked alongside her husband until his early death in 1910 and was honoured as the first woman to be elected to the membership of the Society of Electrical Engineers. Although W. A. Ayrton died eleven years before Michael's birth, it is perhaps no coincidence that one of his research projects was devoted to solving the technical problems posed by the Magic Mirrors of China and Japan. In contrast, their daughter Barbara became a leading Socialist politician, Chairman of the Labour Party and M.P. for North Hendon. After Michael's birth in London in 1921, he grew up close to his father, Gerald Gould, the poet, essayist and literary critic who died in 1936. His formal schooling was broken by long periods of serious illness, the causes of which were never fully identified, and his real education began in the literary and political circles of his mother. In 1935, at the age of fourteen, he had decided to put his father's exposition of the facts of life to the test and was removed from school by Gerald Gould to study independently. From then onwards he travelled extensively, studying art, and although his brief

2 The Bull Ring, 1939

attendances at various art schools including Heatherlys were not particularly successful, it was during these years that he first met Henry Moore and Pavel Tchelitchev. Gerald Gould had died when Michael Ayrton was still only fifteen and his later decision to use his mother's surname was inspired only by, in his own words, the fact that 'A' is the first letter in the alphabet and that, thus, an artist whose name begins with 'A' will always be close to the top of the list in mixed exhibitions. For some years, following the example of Utrillo, he signed his works 'Michael Ayrton.g' in deference to his father's name but dropped the 'g' shortly after 1945. Gerald Gould's gentle character and serious ill-health prevented him from making quite the same impact as Barbara Ayrton in the intensely political circle to which they belonged, but Michael Ayrton remained devoted to his father until his death. However, the strength of pro-Republican feeling in the Ayrton-Gould circle during the Spanish Civil War made a deep impression on the boy and he used to tell the story of how, when he travelled to Paris after leaving school, he went on to Barcelona. Lying about his age, he was there briefly during the bombardment of the city by Franco's troops before Barbara Ayrton discovered his whereabouts. To his fury and relief he was sent home, and a political scandal was averted. Indeed this incident was probably the main reason for Barbara Ayrton's decision to send Michael to stay in Vienna with his remarkable cousin Millie.

Amelia Levetus had been employed, when a young girl, as English governess to the Archduke Rudolf of Austria, and afterwards she settled in Vienna. During her long life she became a well-known bluestocking with decidedly left-wing views, and the year Michael Ayrton spent in Vienna away from his home circle enabled him to establish a considerable degree of independence at a very early age. He copied Old Master drawings in the Albertina and bought minor German woodcuts of the early sixteenth century, but above all he developed the deep interest in politics which he retained for the rest of his life. Socialist by inclination, he remained profoundly unsympathetic to dogma in any guise, and his attitude to politics, as to all other problems, was severely realistic. Cousin Millie rashly stayed on in Vienna after the Anschluss and, aged ninety-one, she was shot down in the Ringstrasse when she

3 Jupiter and Antiope, 1940

struck a member of the Waffen SS with her umbrella. Michael Ayrton, for his part, had to content himself with slightly less heroic anti-fascist activities based in London.

For Michael Ayrton the act of writing, as of speaking in public, came easily (possibly too easily for his own good), but to express himself by drawing and painting was for some years a struggle and always required far greater self-discipline. At the time of his arrival at Heatherlys he displayed great precocity in expressing himself verbally and already wrote in a vivid style, but art came less easily to him and he experienced the utmost difficulty in getting anything onto paper. This was in contrast to John Minton who achieved great artistic fluency at an early age, and although Michael Ayrton was brimming full of ideas there was very often little to show for it afterwards. In autumn 1938 Michael Ayrton and John Minton settled in Paris where they both studied under Eugéne Berman. Ayrton also worked occasionally in the studio of Giorgio de Chirico, and in Spring 1939 they were joined by Michael Middleton, sharing two small studios in Montparnasse between them. Shortly afterwards, the three painters travelled south to Les Baux and spent much of the summer painting close to the ruins of the Renaissance chateau there. Although its inaccessibility made it simple, if not primitive, Les Beaux also hosted a music school in the ruins, run by Diaghilev's conductor, Pierre Monteux, and during 1938-39 music and the ballet occupied a central position in both Michael Ayrton's and John Minton's lives. The exciting stage effects of Jean Girardoux's *Ondine* staged in Paris then had made a deep impact on them, and during 1939 they collaborated on a series of designs for a hypothetical production of Purcell's *Dido and Aeneas.* In these designs the dominant personality must have been John Minton and the importance of the contribution made by him in the formation of Michael Ayrton's early style has not always been fully recognized.

Soon after war was declared Michael Ayrton returned to London and exhibited for the first time, at the Zwemmer Gallery. Apart from the early paintings shown then, he was also working on imaginary theatrical productions, and after these had come to the attention of Sir Hugh Walpole he was recommended to Sir John Gielgud who commissioned the designs for a new production of *Macbeth*. Michael Ayrton asked to be allowed to collaborate with John Minton, and, after both had been called up in 1941, the designs and the supervision of their execution were carried out during their 48-hour leaves. The delayed production was

4 Michael Ayrton with John Minton, 1941

5 The Maskmaker, 1940

6 Strolling Players, 1940

7 *War Fortification, 1940*

finally staged in 1942, and in the winter of the same year Michael Ayrton was invalided out of the R.A.F. Indeed, his service in the R.A.F. had not been without its problems, and he developed sudden, very high fevers with delirium which equally suddenly abated. The R.A.F. attitude was that he was a malingerer and he was interviewed by an R.A.F. psychiatrist who suffered from a facial twitch and shuffled money in a drawer while talking to his patient. As Michael Ayrton got to know him better their rôles changed and in his own words,'I nearly cured the psychiatrist's twitch but I couldn't quite get at his dreams!' His physical condition, however, continued to deteriorate and Michael Ayrton returned to civilian life in wartime London. Shortly afterwards, a deep-seated throat abscess was diagnosed by Somerville Hastings and cured by massive Vitamin B treatment, while together with John Minton he held an exhibition of oil paintings and theatrical designs at the Leicester Galleries. On the strength of the success of this, and the designs for *Macbeth,* he obtained a post teaching life drawing and theatre design at Camberwell School of Art (1942-44).

The paintings executed by Michael Ayrton in Paris, such as the *Comedia* which was painted in April-May 1939, are relatively undistinguished in quality though his choice of subject matter, drawn mainly from the world of the circus and the bullring, was to be maintained into the early 1940s. Nevertheless, the rapidity with which he reached artistic maturity was most remarkable and even by *The Bullring,* signed and dated July 1939, his treatment of sculptural form had taken on considerable authority. From the beginning Michael Ayrton's painting style had its roots in the tradition of *disegno* rather than *colore* and in the expressive outlines of his figures lies their impact. In his strongly evocative drawing *The Vase*, signed and dated 29 January 1940, the principal source of this linearity is clearly revealed and the long-term influence of the drawing style of Dürer cannot be overestimated. The forms are built up with patches of close cross-hatching, strongly reminiscent of the techniques of Dürer, and in a series of drawings including *The Maskmaker* (signed and dated February 1940 and now in the Bankfield Museum, Halifax) and *Jupiter and Antiope* (signed and dated February 1940) Michael Ayrton developed his highly personal, bizarre imagery. The growing spatial ambiguities and weird details of his *Juggler in Sequins* (1940) lead on to the *Fête Champêtre* painted from July 1940 in which the strong elements of Surrealism and the debt to Picasso and de Chirico are self-evident. This phase in Michael Ayrton's evolution as a painter also provides the point of departure for his earliest theatrical designs and the gouache drawing *War Fortification* (1940), now in the Victoria and Albert Museum, is a prelude to the backdrops for *Macbeth.*

Unfortunately, few of Michael Ayrton's paintings of the early 1940s can now be traced, but the deeply moving naked figure of *Orpheus* in a bleak landscape, executed in pale pinks, greys and blues in May 1941, has something of the taut outlines and emotional intensity of Ferrarese Quattrocento painting. Not a self-portrait, this painting nevertheless takes over some of the elements of narcissism

8 Fête Champêtre, 1940/41

9 The Juggler in Sequins, 1940

to be found in the self-portraits of Dürer and in particular it brings to mind the drawn *Self-Portrait* by Dürer now in the Pierpoint Morgan Library, New York. During 1941 Michael Ayrton's friendship with the composer Cecil Gray and the mystical circle including Barnet Stross, Freda Cavell, James Laver and above all Margery Livingstone, and its links with Alistair Crowley, began to take on a new significance. It was to lead to Michael Ayrton's decorations for the edition of Cecil Gray's play *Gilles de Rais* published in 1945, but echoes of more deeply disturbing elements are reflected in his own short story *A Performance of Saint Joan* published many years later in *Fabrications* (1972). The obsessional themes of *The Temptation of St. Anthony* and the *Agony in the Garden* commencing in 1941/42 are in certain respects on a line totally separate from the Gilles de Rais ideas. Formed entirely in his own head, Michael Ayrton saw St Anthony as himself either then or as he would be, and nobody was allowed emotionally close to it. Undoubtedly there was an element of catharsis in this process and in 1943, at the time of painting *Sleeper in Flight,* he was deeply concerned with death, as is revealed by his poems published then in *Encounter.* Towards the end of the run of *Macbeth* in London in 1942 Michael Ayrton began the close relationship with Joan Walsh which was to last until 1949 and at first they lived in the same flat as Barbara Ayrton in 67 Belsize Park Gardens. *The Temptation of St Anthony* was begun in 1942 in his studio there and Michael Ayrton painted his own arms and legs to shadow the muscles and emphasize the veins before executing, with the use of mirrors, the figure of St Anthony from himself. This was exceptional, marking his deep emotional relationship with St Anthony, since Michael Ayrton very rarely worked from posed models. He preferred spontaneous poses and was thus, for example, a very reluctant portraitist unless he was particularly interested in the sitter as a person. In the middle of painting *The Temptation of St Anthony* however, Michael and Joan moved into a large studio in 4 All Souls Place alongside the B.B.C. and, in many respects the finest of the series, *Skull Vision* was painted there early in 1943, though he continued to be fascinated by the *St Anthony* theme proper until 1946.

10 Design for Macbeth, *Act II, 1941*

11 Design for Malcolm, in Macbeth *Act I, Sc. 2, 1941*

12 Set design for Macbeth, Act III, Sc. 9, Dunsinane, 1941

(facing page) *13 Orpheus, 1941*

michael ayrton

14 St Anthony No. XXI, 1942

15 Study for St Anthony, 1942

16 Study for the Child in Temptation of St Anthony, 1942

17 St Anthony's Shadow, 2nd Series IV, 1945

18 Temptation of St Anthony, 1942/43

On his arrival at Camberwell School of Art in 1942, aged twenty-one, Michael Ayrton felt too youthful in appearance so he grew a beard to endow himself with an air of authority, but his months of copying prints and drawings in the Albertina in Vienna stood him in excellent stead and he was able to demonstrate the drawing techniques of different masters and periods with an alarming facility. In retrospect, Camberwell made relatively little impact on Michael Ayrton's stylistic evolution and much more influential was the summer of 1943 which was partly spent with the Mallesons at the White House, Little Missenden. *Mrs Chittenden gathering Cabbages* was painted in London in July 1943 from sketches made in the garden of Richard Gorer (the purchaser of *The Temptation of St Anthony)* and of his housekeeper, but *The Sleeper in Flight* (formerly in the collection of Robert Donat) was painted in Little Missenden in the high summer of 1943 and marks the maturity of the next phase of Michael Ayrton's painting style. Whereas the German masters of the early Sixteenth Century, and above all, Dürer, Grünewald, and Pieter Bruegel the Elder provided the principal sources of inspiration for the St Anthony sequence, a new lyricism and delight in the English landscape now becomes apparent. In *The Sleeper in Flight* the figure, based on the youthful Nicholas Malleson, hovers above a delicate landscape reminiscent of Paul Nash, while in the contemporaneous *Joan in the Fields* these sources are wedded to the Sudbury portraits of Gainsborough. *Mrs. Chittenden* is remarkable for the freshness of Michael Ayrton's palette and, after the black introspection of the St Anthony sequence, for the sense of fresh air and rapprochement with nature. *Joan in the Fields* was completed in the All Souls' Place studio in autumn 1943 and an increasing Romanticism is to be seen in *Il Desdichado* painted in April-May 1944. Although his personal contacts with Paul Nash were relatively slight, Nash's late landscapes made a deep impression on Michael Ayrton at this time and their influence can be traced as late as 1946 and the *Track to Chanctonbury.* In this respect, however, Michael Ayrton belonged to a widely diffused movement in English painting of the mid 1940s away from Surrealist imagery towards an immediate response to nature. The summer heat of *Earthbound* (1944) with the sleeping figure growing as much from the soil as the standing wheat and the stooks beyond, belongs to the same world as that then painted by Paul Nash, Victor Pasmore and Ivor Hitchens.

Much of the summer of 1944 was again spent in

19 Skull Vision, 1943

Little Missenden, with visits to both the Nashs and the studio of Ivor Hitchens, but the summer holidays of 1945 and 1946 were spent in Pembrokeshire where Michael Ayrton was in contact with Graham Sutherland. Although the imagery of *Tree Forms* and *Convulsive Landscape* (both 1944) owe much to Paul Nash and his *Monster Field*, the animal quality of the tree forms is more explicit and the sinister gives way to an erotic *frisson*. During the late 1930s Graham Sutherland steadily rejected his indebtedness to Samuel Palmer, but the magic of the Shoreham watercolours was re-evoked by Michael Ayrton briefly at the beginning of 1945 in a brilliant series of drawings including *Apple Trees by Moonlight* (executed in April 1945 but present whereabouts unknown), and by the summer he was entering a new phase with a series of exuberant landscapes exploiting tangled roots and branches. Revealing parallels can be drawn between the excellent *Entrance to a Wood* painted in May-June 1945 (Arts Council of Great Britain) and the vigorous landscapes of 'the Roberts' (Colquhoun and MacBryde) who had briefly shared a studio with John Minton early in the war. Indeed that studio was a popular rendezvous for a group of young artists and writers including Keith Vaughan, Dylan Thomas, Prunella Clough and John Craxton as well

20 Sketch for Skull Vision, 1943

as Michael Ayrton. The *Entrance to a Wood* was painted at the same time as the *Great Yews of Odstock* in the house of Michael Ayrton's godmother at Coome Bissett outside Salisbury, and *The Honeysuckle Picker* (present whereabouts unknown) belongs to the same group of landscapes, while *Winter Stream* (March, 1945; Aberdeen Art Gallery) and the large version of *Gethsemane* are a prelude to this phase.

21 Joan in the Fields, 1943

22 The Earthbound, 1944

23 The Sleeper in Flight, 1943

24 Convulsive Landscape, 1944

25 Tree Forms, 1944

26 *Apple Trees by Moonlight, 1945*

27 Mrs Chittenden, 1943
28 Entrance to a Wood, 1945

29 Winter Stream, 1945

In 1944, at the age of twenty-three, Ayrton threw up his teaching job at Camberwell and began to write seriously, taking over from John Piper as art critic to *The Spectator.* He also began to illustrate books and continued to design for the theatre, though his interests in the latter field (until the death of his close friend, the composer Constant Lambert, in 1951) were now concentrated on ballet. Lambert had moved into All Souls' Place during the first half of 1944 and was one of the few real influences on Michael Ayrton. Their friendship was close and in his *Sketch for a Portrait* published in *Golden Sections* (1952) Michael Ayrton has recounted with a certain sadistic satisfaction their visit to Stoke-on-Trent and the Sadlers Wells Ballet Company early in 1944:

> 'Sunday in Stoke-on-Trent, in Hanley, Burslem, Tunstall, Longton, Newcastle-under-Lyme and in Etruria is an austere celebration. Constant and I began by being forbidden to play billiards in his hotel. The late morning was spent in a visit to the Great Tip, where a million broken cups and saucers make for rats a porcelain Chicago. Lunch must be passed over in silence. In later years the subject of Sunday lunch in Hanley was tacitly passed over by us both. But the grey and glacial public gardens of Stoke and the walk by the canal will always remain with me as one of the most macabre excursions ever shared by a painter and a musician. Any

30 The White Country, 1945

trifling pessimism apparent in Constant's music or my painting since that date may be directly attributed to the effect of that hideous peregrination'.

The impact of that visit upon the innocents can, nevertheless, be felt strongly in the strangely moving painting, *The White Country* (1945), with its moonlit vista of industrial desolation, and again, more explicitly in *The Tip, Hanley* painted in 1946.

During the two years for which Michael Ayrton held the post of art critic of *The Spectator* he rapidly emerged as a writer of considerable distinction and his scathing essay on Picasso entitled *The Master of Pastiche,* which was published in 1944 (reprinted in *The Rudiments of Paradise,* 1971), earned him early notoriety. Apart from his father and the Ayrtons, the literary background of Michael Ayrton was conducive to an early development, and as the nephew of Israel Zangwill a certain forthrightness and taste for satire was to be expected. The first mark that Michael Ayrton had left on the literary scene was when, as an incontinent baby, he had been placed in the lap of John Galsworthy, and in later life his radical leanings did not allow him any regret for the incident. His literary background notwithstanding, Michael Ayrton's school career was totally undistinguished and the influence of Cousin Millie in Vienna was a significant factor in his later development as an art critic and art historian. The *Master of Pastiche* remains a formidable piece of critical analysis for a twenty-three year old and many passages in *British Drawings* written the following year (published 1946) are equally memorable. Much of the latter was written, with Joan's encouragement, in Cecil Gray's windmill at West Chillington near Pulborough, but although regular art criticism was highly stimulating, Michael Ayrton soon came to the conclusion that he could not combine it with painting and after two years he resigned from *The Spectator.* His desire for independence was no doubt encouraged by the friendship with Wyndham Lewis which he had first struck up in 1945 and which culminated in the close co-operation of the two artists after Lewis went blind in 1951 and was forced to abandon painting. As an art critic, Wyndham Lewis was deeply impressed by Michael Ayrton, and his last written work was the introduction to Ayrton's *Golden Sections,* published in 1957.

In *Round Britain Quiz,* and as almost the first T.V. art pundit, his urbanity and quite extraordinary range of knowledge made him an enormous success with the British general public. The British critic, however, hates versatility and, perversely, the popular acclaim he received then as a broadcaster undermined the critical appreciation that his paintings were beginning to receive. James Laver's study of 1948 remains the only book to be devoted to Michael Ayrton's paintings. Politics left another mark in the sequence of industrial and related subjects which he began in response to the election of the Socialist Government in 1945. In 1942 he had briefly explored the subject matter of the *Chainmakers* and in late 1945 he returned to it in *The Forge* (dated November 1945-February 1946). *The Tip, Hanley* (1946) belongs to this sequence (Sir Barnett Stross was the Labour MP for Stoke and a close friend from the early 1940s), while the Thames docks scenes executed during the grim winter of 1946-7 are in direct succession. A different line of development, springing more from the exuberant landscapes of 1945, are the spirited sea pieces including *Headland, Storm Blowing Up* (formerly in the collection of Samuel Courtauld), *Dark Sea* (Ferens Art Gallery, Hull), and *Skokholm Island,* all painted in 1946.

31 The Forge, 1945/46

32 The Tip, Hanley, 1946

Despite their success, the theatrical designs of Michael Ayrton, like his book illustrations before the Archilochos Etchings, remained detached from the main development of his style, and the pen and ink and bodycolour drawings for *Macbeth* established the pattern for his later productions. The designs for *Macbeth,* executed in partnership with John Minton, were not the product of their first collaboration (designs for Purcell's 'Dido and Aeneas' had been executed by them in 1939) but it was the joint character of that partnership which established the basic style for the later designs. Often employing

33 Headland, Storm Blowing Up, 1946

brilliant colour contrasts,as in certain of the costume designs for Constant Lambert's production of Purcell's 'Fairy Queen', the easy flow of the pen strokes and the highly effective use of broad areas of flat body colour provide a close parallel to the book illustrations executed in the early 1940s. A group of the *Macbeth* designs was subsequently reissued as book illustrations in the Folio Society second edition. Michael Ayrton's own account of the designs for the *Fairy Queen,* included in the souvenir produced by the Sadlers Wells Company, is particularly informative. Of his book illustrations, the series of twenty-nine drawings executed in 1941 for an edition of Paul Verlaine's *Fêtes Galantes* remains unpublished and one must turn instead to the delicate pen drawing produced as the frontispiece to David Cleghorn Thompson's *The Hidden Path* published in 1943. Amongst the most successful book illustrations are the wiry pen drawings decorating Cecil Gray's *Gilles de Rais* (1945) and Thomas Nashe's masque *Summer's Last Will and Testament* (music by Constant Lambert, 1946), and the wash drawings for John Arlott's *Clausentum* (1946). The later illustrations, with the exception of those for

34 Haytor Quarry, 1947

35 Ebb Tide

36 The Quay, Isle of Dogs, 1947

his own *Tittivulus or the Verbiage Collector* (1953), generally lack the tautness of the best early drawings and it is not entirely surprising that Michael Ayrton lost interest in illustrating books during the 1950s. Subsequent book illustrations—in particular, the splendid series of two-colour lithographs for Apuleius's *The Golden Ass* (1960), and the erotic etchings for Paul Verlaine's *Femmes, Hombres* (1971)—tend to be lateral developments from his sculptural preoccupations, or relatively unimportant productions. Similarly, during the late 1940s, Michael Ayrton lost interest in theatrical design work and gave it up entirely after Constant Lambert's death in August 1951.

37 *Thames Foreshore, 1947*

Far more important, however, was Michael Ayrton's first visit to Italy in 1947 which sparked off his deep love for the Italian Renaissance. Throughout the war years the collections of the National Gallery and the Victoria and Albert Museum had been away in store and the impact of the Italian Renaissance and Italian light after almost six years in wartime England cannot be over-emphasized. During this relatively short visit he mainly looked, and it was not until the winter of 1947-48 that the direct influence of the Quattrocento painters, in particular Masaccio and Piero della Francesca, began to be felt in his paintings. Michael Ayrton and Joan left London in mid-July and, travelling via Rome and Naples, reached Ischia by the end of the month, to be joined soon afterwards by Constant Lambert. Contact was maintained with Norman Douglas, then staying on Capri, and apart from a large number of sketches, Michael Ayrton painted a few small canvasses and panels. Their visas ran out and after a frantic bus journey from Naples to Genoa a sympathetic official renewed them and they travelled back to Florence. During a stay of a month in the Villa Diana Michael Ayrton developed the Ischia material and first saw the Barna da Siena frescoes in the Collegiata in San Gimignano. His short essay, *La Madonna del Parto,* an account of his first visit to the Cemetery Chapel at Monterchi, written in 1947, conveys superbly the response to Italy of Michael Ayrton the writer as well as Michael Ayrton the painter, and although his essay on Barna da Siena, entitled *The Terrible Survivor,* was not published until 1956, its genesis began in 1947. Michael Ayrton's normal working practice was to execute drawings and perhaps a few small oils (about 10in. by 14in.) during his journeys and to work on his major compositions in his London studio after his return.

38 *Design for Head of Mephistopheles, in* The Damnation of Faust, *1947*

39 Design for Backdrop 'The Wood' for The Fairy Queen, *Act I, 1946*

40 Mephistopheles, in The Damnation of Faust, *1947*

41 Design for Oberon, in The Fairy Queen, *1946*

The final synthesis of ideas gained during this stay in Italy was achieved in London in his monumental *Afternoon on Ischia*, now in the collection of the British Council, and the less well-known *Conversation* (also of late 1947). *Approaching Thunder* (Exeter College, Oxford) was painted in All Souls Place in February 1948 from the Ischia sketches, while the spirited two-colour lithograph *Conversation in the Wind* published by the Redfern Gallery dates from the same time, together with *The Sudden Shower* (private collection, New York). These Italian paintings were exhibited at the Hanover Gallery in 1948 under the title *Italian Journey.* Never a natural painter, Michael Ayrton developed into one of distinction by the late 1940s through the dogged determination and singlemindedness which characterised so many of his enterprises, but it is nevertheless important to recognize the limitations which he imposed upon himself. These are summarised in the statement which he made in June 1949 for the retrospective exhibition organised by the Wakefield City Art Gallery:

> 'Iam not really gifted with 'painterliness' in the French sense, a shortcoming which critics have not been slow to point out. Paint texture and handling of the material interests me no more than it interested Hugo van de Goes or Antonello da Messina and indeed the overriding influence upon my recent work of artists who were primarily concerned with fresco (wherein the presentation of the image was the first concern and the surface of the artifact was, apart from the achievement of a smooth and proper finish of relatively small concern), has inevitably reduced my interest in 'handling' to a minimum. A critic has accused me of applying paint like a skilful housepainter and I take no exception to this since I am concerned with rendering images as simply and as austerely as I am able and not in producing delectable or succulent pictorial dishes. The work of sculptors was the most potent single influence on painting during the formative years of the Renaissance; Giotto owed much to the Pisani, Masaccio much to Donatello, and Leonardo's master was Verrocchio. Sculptural form realised pictorially by drawing was at the centre of the Renaissance and is the significant factor in the 19th century art of Ingres, Degas and Seurat and in the 20th century art of Juan Gris and Wyndham Lewis. It is also the chief virtue of the Cubists. Herein my main interest lies. I would go further and maintain that this astringent form of art is needed to-day to balance the powerful influence which Impressionism, Expressionism, Neo-Impressionism and so on, still combine to exercise upon contemporary painting. I would not speak against this alternative approach to painting which has proved so magnificently expressive in the hands of great masters since Titian and which has given us, Velasquez, Turner and Renoir, to name three disparate but undeniably supreme artists, I merely say that this 'painterly' approach is not an indispensable one to the making of pictures and furthermore it has been the source of a great deal of superficial and meretricious visual culinary. I aim, rashly enough, at the creation of monumental images as timeless, as impersonal and yet as human as those of Piero della Francesca and Giovanni Pisano, as filled with secret communication and unspecified drama as those of Masaccio. This grandiose ambition will fail, probably in every particular, but it is to me preferable to the current and general desire of painters to express an original opinion about the critical situation in which the artist finds himself to-day, and which painters often tend to express by loading their canvas with the bloated flesh of emotionalism, the succulent fat of 'paint quality,' the alluring vestment of rugose and romantic obscurity or the glittering draperies of undisciplined light. I should not dream of denying that I have at times succumbed to all these tempting mannerisms, I merely deny that they are significant. I strive to concern myself with the less perishable bones.'

42 The House, Ischia, 1947

The rejection of painterly qualities and the emphasis on sculptural forms are the strongest elements in Michael Ayrton's paintings of the late 1940s, and the larger Italian paintings, when exhibited at the Redfern Gallery in 1949, received a critical acclaim which had previously eluded him.

In summer 1948 Michael Ayrton returned to Italy, but this time in the company of Nigel and Elisabeth Balchin, and apart from continuing his study of Italian Renaissance art in Tuscany he made a

43 Afternoon on Ischia, 1947
44 Approaching Thunder, 1948

45 The Sudden Shower, ?1947/48

46 Conversation in the Wind, 1948

considerable number of sketches and painted the small *Portrait of Norman Douglas* now in Sheffield. The latter, executed in the Hotel Berchielli in Florence, follows in the tradition of the small Ischia paintings of 1946, but the much larger and more ambitious *Full Moon* in the collection of the British Council was painted in London during the following winter although the scene was based on one that Michael Ayrton had seen in Arezzo that summer. The small boy on the park bench beside a fat man teases a cat while two other children point and exchange confidences. In the context of the painting it is left ambiguous as to whether it is the fat man or the pair of lovers further along the path who are the source of the children's disapprobation, and the baleful moonlight strengthens the slightly sinister note of the scene. Early in 1949 Michael Ayrton returned to Italy with Elisabeth Balchin for some six weeks and for the first time witnessed the preparation of the vineyards. The tying up of the vines and the pruning of the trees over which they are trained provided the source of inspiration for the series of Tuscan vineyard scenes including *Vines before Easter* and *The Passion of the Vine,* both 1949. Set in the Tuscan hills with the figures and vines silhouetted against the clear skies and crisp landscapes, the Vine Sequence reveals an economy of means and austerity which Michael Ayrton rejected in such later summer and autumn landscapes as *Olives and Vines* (1947) and *Oxen under the Vine* (1950). In these patches of delicate

47 Full Moon, 1948/49

48 Portrait of Norman Douglas, 1948

colour are laid on freely with precisely the painterly effects he in theory eschewed. More closely related in treatment to the Vine Sequence is *The Shepherd* of 1950 which was a theme Michael Ayrton often returned to later in his career. The image of the lamb carried across the shoulders of the young shepherd is based on the Antique sculpture in the Uffizi rather than the Moscophoros which was to become so important to him in the mid-1950s.

50 Good Friday, 1949

49 The Shepherd, 1949

51 The White Cat, 1949

52 *Olives and Vines, 1947*

53 *Vines before Easter, 1949*

54 The Passion of the Vine (VI), 1949

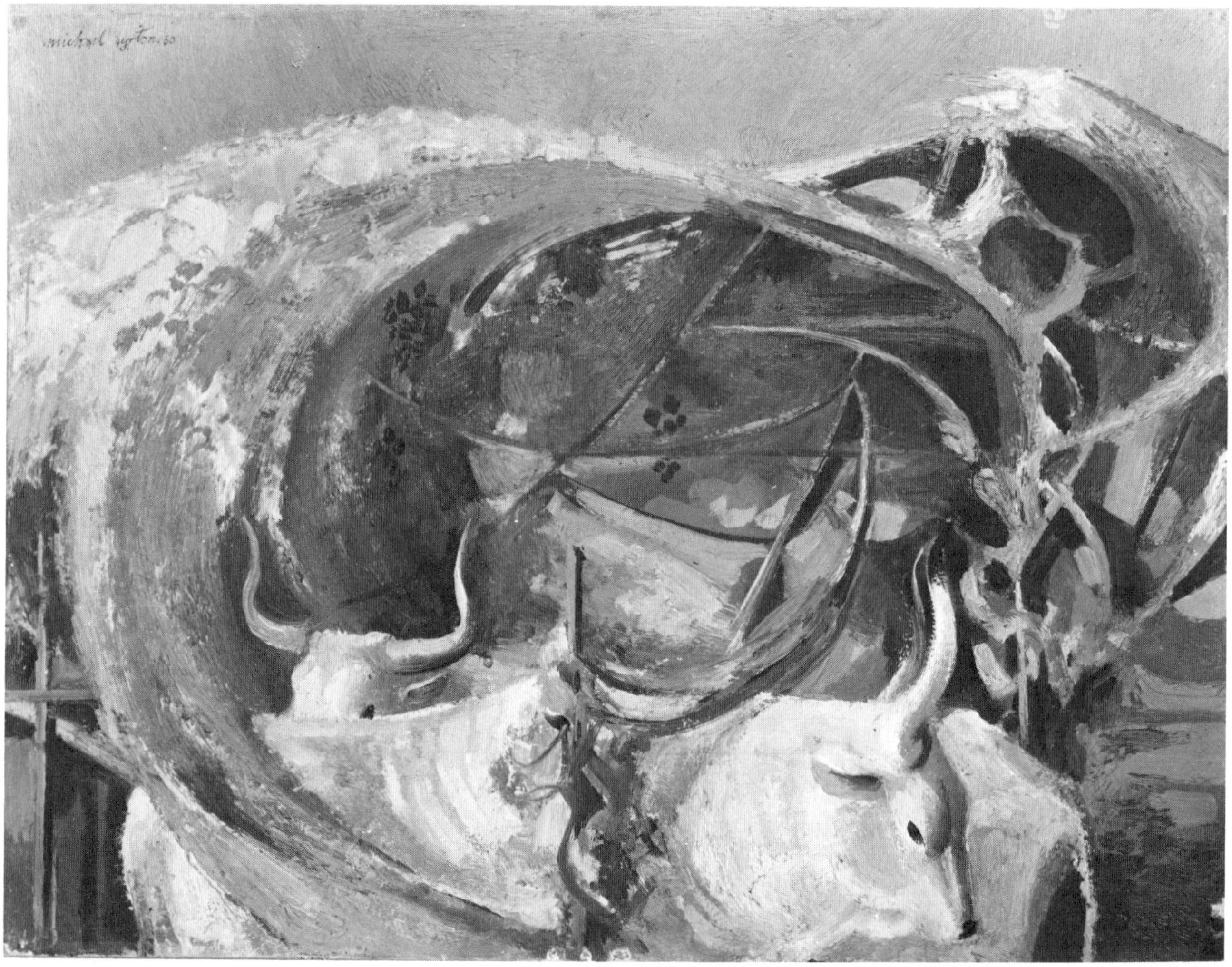

55 Oxen under the Vine, 1950

Soon after Michael Ayrton's return to London in April 1949 his relationship with Joan finally broke down and he was joined by Elisabeth Balchin. Life at All Souls Place had been changing steadily as he spent longer periods abroad and the close proximity to the B.B.C. proved something of a mixed blessing despite the unending source of diversions which it provided. Constant Lambert had moved into the lower floor early in 1944, but on his marriage to Isobel Delmer in October 1947 he moved into Albany Street and his departure was felt keenly. This was despite the many visitors, which included Edith Sitwell, Humphrey Searle, Rose Macaulay and Dylan Thomas. Notwithstanding a continuing friendship with Aneurin Bevan and Barnett Stross, Michael Ayrton's political interests retreated further into the background, to the intense disappointment of his mother, and her final illness coincided with the break-up of his relationship with Joan. Indeed the period around 1950 was a time of major changes, the most important being his marriage in that yearto Elisabeth Balchin and his acquisition of three stepdaughters. Until early in 1951 they continued to use All Souls Place as their centre when in London, and they moved into Bradfields in time for Christmas of that year. For the Festival of Britain, Michael designed *The Elements as the Sources of Power* which was enlarged photographically into a mural 120 feet in width.

The preparation of the large exhibitions of paintings shown in Milan and Rome in 1950 and his extended stays in both cities led to Michael Ayrton's taking a considerable interest in Italian town life as well as in the Tuscan landscape. Street urchins playing, in paintings such as *The Wall, Version I* (1950), had attracted him before, but much more specific was the new interest he now took in the intense life of the Rome slums of Trastevere. In the large oil painting *Roman Window* (1950) in the Swindon Art Gallery, the dismal squalor portrayed is transmuted by the brilliant light and the open window reveals a scene of earthy reality. Nevertheless, Michael Ayrton was also aware of the other side of life in Trastevere, the crime and viciousness bred of poverty, and a distinctly sinister note is struck by *The Captive Seven,* painted in London after his return. The pencil study for this ambitious composition, made in September 1950, reveals the clear formal structure, rooted in the disciplines of the Italian Renaissance, as well as the construction of the both disturbed and disturbing group of figures enclosed within the semicircle. As with the few other paintings by Michael Ayrton for which true preparatory studies were executed, the finished painting is marked by an appreciable toning down of the biting observation characteristic of the drawing. In *Mirror Image,* painted in October 1950, with the subtle use of a mirror and painting within the painting, the ideas of the Trastevere paintings are translated to London, as is also true of the painting ambiguously entitled *The Sisters* (executed in 1951). The brush drawing

56 *Roman Window, 1950*

57 *Ayrton in his Studio in All Souls Place, London, 1950*

of the latter composition is again considerably more sinister, although the final touch of a cigarette in the hand of the girl seated on the bed adds the additional note of decadence which was avoided in the wistful painting *Girl on a Bed* of 1951-53. The secret life of Trastevere, either in the original or translated to London, clearly intrigued Michael Ayrton in 1951, and the refined eroticism of the pen and wash drawing *Trio* executed in October 1951 is a further development of this theme, while a lateral development of the Trastevere interiors is the *Figure in a Yellow Room* (1951).

58 The Elements as the Sources of Power, 1951

59 Girl on a Bed, 1951/53

60 Trio, 1951

61 The Sisters, 1951

62 The Sisters, 1951

63 The Captive Seven, 1950

64 Study for The Captive Seven, 1950

65 *John Gielgud as Macbeth, 1942*

66 *Figure in a Yellow Room, 1951*

As has been noted previously, Michael Ayrton was a reluctant portraitist unless he was particularly interested by the sitter. His sympathetic *Portrait of John Minton* executed in 1941 is stylistically close to *Orpheus* and is executed in the same range of delicate colours, while the *Portrait of Sir John Gielgud as 'Macbeth'* painted in August 1942 draws together his oil paintings and his theatrical designs although it is not particularly distinguished in quality. Much more important for his later development is the *Portrait with Pears* of 1944 which has been identified as a portrait of the caretaker of the Camberwell School of Art. In many respects the paint handling and interest in sculptural form anticipates Michael Ayrton's Italian paintings and the visual pun between the shape of the lady and the fruit is unmistakable. The brush and ink *Self Portrait* executed in 1946 is remarkably cool and detached and as such illustrates well the ruthlessness and objectivity that Michael Ayrton cultivated during his earlier years and tended to conceal behind his extrovert life style. It is in many respects an extraordinarily revealing document, as if momentarily a mask had been dropped, and the finest of his later brush portrait drawings from *Dylan Thomas* (1947) to *William Golding* (1965) are amongst the best post-war portrait drawings produced in England. Unusual for Michael Ayrton is the penetrating pencil drawing of *Sir William Walton* executed in Italy in May 1948 with the forms built up from a network of continuous lines. Varying in pressure, the pencil moves over the sheet of paper in a single continuous movement of great virtuosity. The drawing is now in the collection of the Boston Museum of Fine Arts and the direct impact of Wyndham Lewis is clear. Michael Ayrton, despite his strong objections to the political views of Wyndham Lewis, put forward Lewis's claims to be regarded as the finest portrait draughtsman of his time in *British Drawings* (1946) but it was not until 1953 that a close relationship was forged, two years after Lewis went blind. In the few years before Lewis's death in 1957, Michael Ayrton executed a number of portrait drawings and the memorable oil (now in an English private collection). Wyndham Lewis's sharply chiselled features under the green celluloid eyeshade he adopted in his blindness, his expressive hands and his vorticist handkerchief, together with the formidable ash-tray which had fascinated Michael Ayrton when they had first met in 1945, constitute an image of great potency. In later years Michael Ayrton rarely lost an opportunity of drawing people who had particularly interested him, or for whom he felt a special sympathy, and

67 Portrait of John Minton, 1941

although the relatively formal drawings of Henry Moore and Graham Sutherland do not rise to quite the same heights as his earlier portrait drawings, they remained an important element in his artistic production.

In the course of a visit to Paris during the first half of 1950, Michael Ayrton told his family that he wished to sculpt and much of that stay was spent in long conversations with Giacometti in his studio. He felt that he could not sculpt in London because he needed a great deal of additional space, and he was equally aware of the necessity of moving away from Broadcasting House and its time-consuming diversions. To this end the Ayrtons rented a house in Wiltshire for the summer of 1950, as much to see whether they could live happily in the country as to obtain the peace he needed to work. This stay proved so successful that they looked for a house to buy in Wiltshire, to no avail, and soon afterwards purchased Bradfields in Essex from Sir Francis Meynell. Michael Ayrton had bathed in the pond there as a child and had happy recollections of the

68 *Portrait with Pears, 1944*

house, but a particular attraction was the large barn adjacent, with a concrete floor, which would provide a ready-made sculptor's studio.

In retrospect, the move from London to the Essex farmhouse of Bradfields near Toppesfield was inevitable, but in 1951 for the highly successful broadcaster and public personality to exile himself deliberately in an inaccessible corner of Essex was interpreted by some as just another manifestation of his perverseness. During the mid-1940s he had executed a series of landscapes showing a deep feeling for the English Neo-Romantic revival and for the especial qualities of English and Welsh landscape, and he had later grown to understand the Tuscan landscape with equal sensitivity. *Winter,* however, painted in 1950 with the curling bare branches of the trees outside the window silhouetted against the winter landscape and a skull on the window-sill comes as something of a surprise. In his yearning to expand into three dimensions, space in which to breathe freely became increasingly important, and the move to Essex in 1951 may be seen more as a symptom than as the cause of his development in the early 1960s. At the same time, he turned again to the theme of *The Shepherds* with the monumental paintings such as that in the City Art Gallery, Manchester, (1951), a large group of related drawings and the excellent lithograph (1951). In the tiny *Cornfield* (1951), Michael Ayrton revelled in the ripe wheat of an English summer, but it is perhaps in his paintings involving cats in the early 1950s that the new serenity most clearly prevails. The Ayrton household has never been complete without a large complement of cats, a love shared with Constant Lambert at All Souls Place. The painting *The White Cat* of 1949 leads on to the enchanting brush and wash drawing, again of two girls holding a cat, of February 1952, entitled *The Unforeseen Event.* Three step-children in a large Essex garden inspired a series of outdoor scenes including *Figures with a Tortoise* (1953) and, in a slightly different vein, the sequence of still life paintings from the *Green Still Life* of 1953 (given by Lord Eccles to the British Embassy in Rome) to the sumptuous *Still Life at Christmas* (1953-54; The Arts Council of Great Britain). The genesis of the latter was described in detail by Michael Ayrton in his essay *The Glutton Eye* published in 1954, and his deepening involvement in the cyclical processes of nature 'as the orange shout of the narrow flower, lying on its neighbour's heavy child, sang from the green shadows'.

69 Portrait of Dylan Thomas, 1947

70 Portrait of William Walton, 1948

71 Portrait of Norman Douglas, 1948

72 Portrait of Constant Lambert, 1951

73 *Portrait of William Walton, II, 1948*

74 *Portrait of Wyndham Lewis, 1953*

75 *Portrait of Graham Sutherland, 1967*

76 *Portrait of William Golding, 1965*

77 *The Unforeseen Event (The Improbable Circumstance), 1952*

78 *Winter, 1950*

79 The Shepherds, 1951

80 The Shepherds, 1951

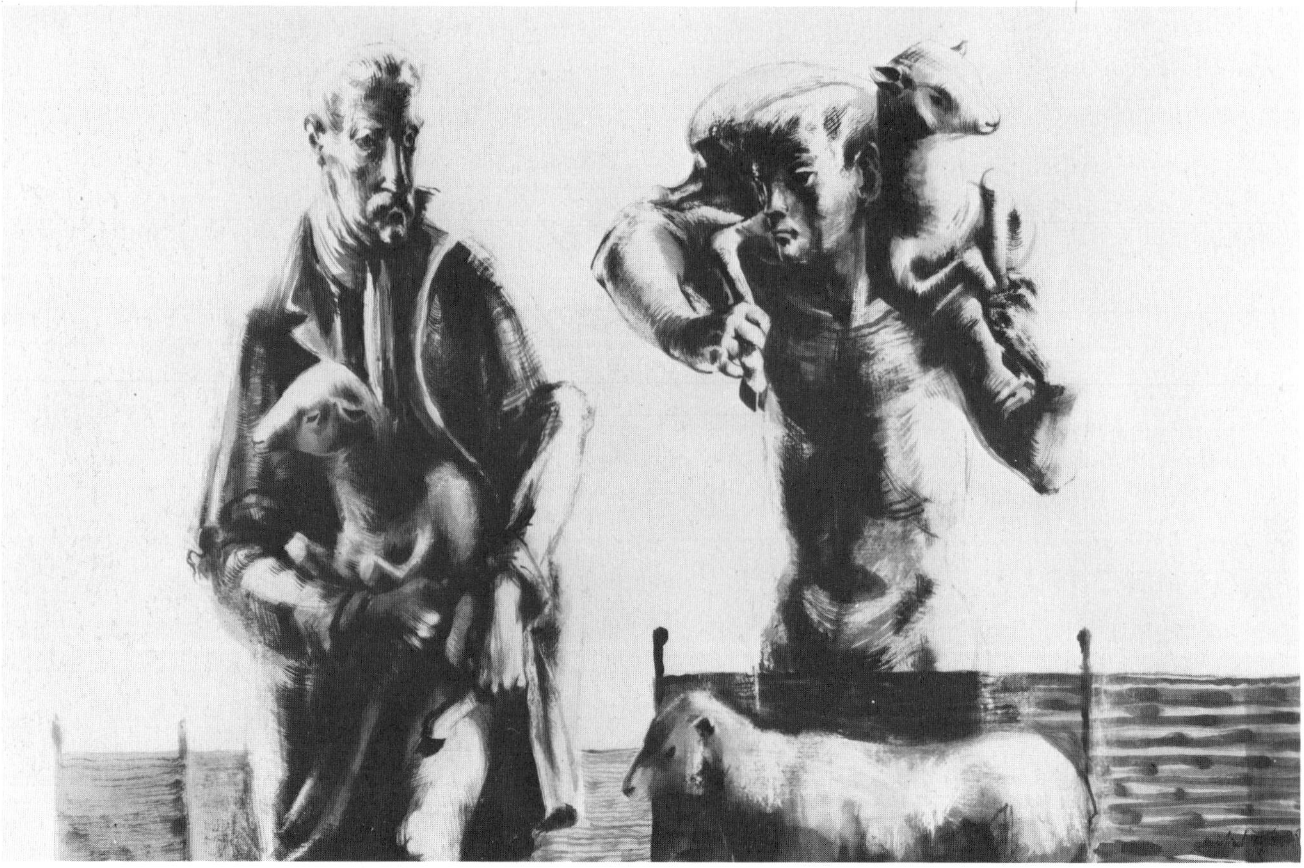

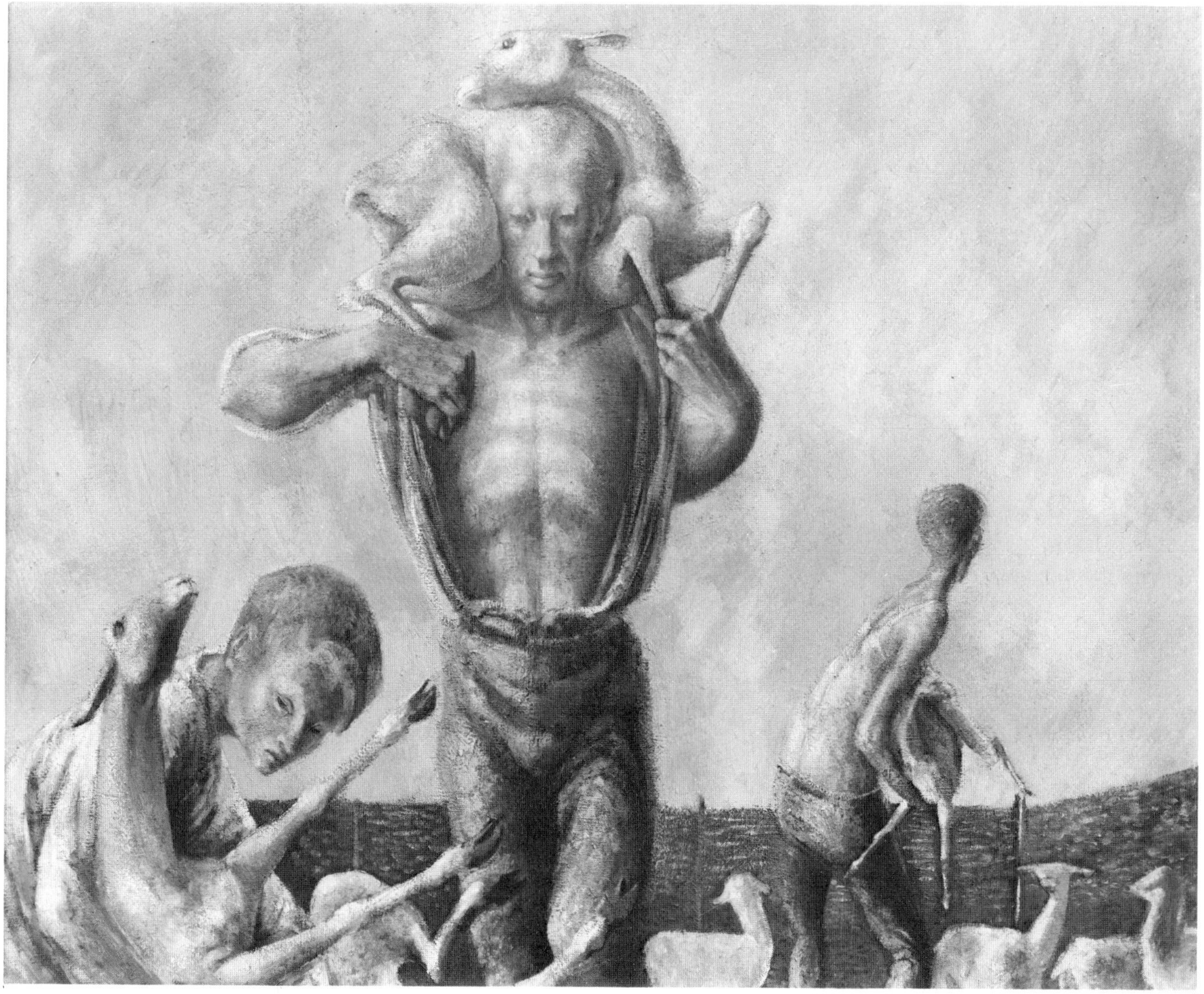

81 *The Shepherds Large II, 1951*

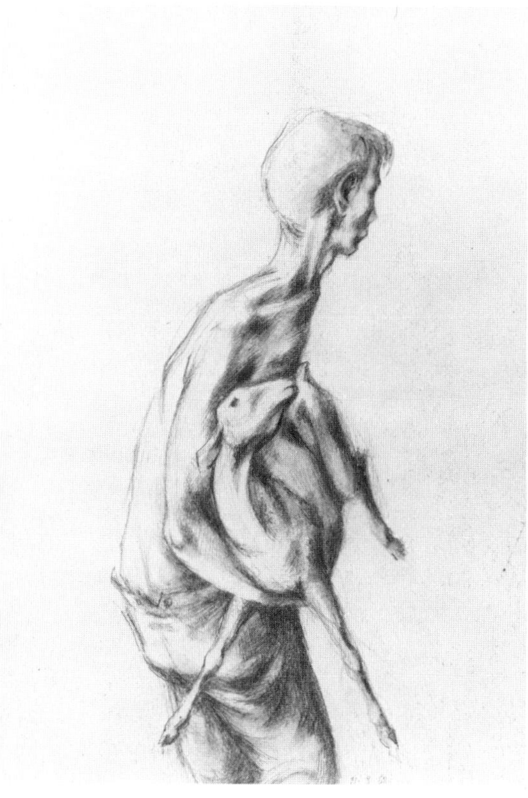

82 *Study of a Shepherd, 1952*

83 *The Shepherd, 1951*

84 Green Still Life, 1953

85 Still Life at Christmas, 1953/54

86 *Still Life — Carnations and Lemons on a Table, 1953*

87 *Traveller's Joy, 1953*

88 Blind Girl with Sculpture, 1952/53

The early 1950s were indeed times of change, but as far as his paintings were concerned, it was change with little real sense of direction, if not regression. In the late 1940s he had appeared to reject popularity by painting in a style which seemed to be austere and which was often deliberately unappealing, while continuing to attract widespread public attention as a controversial critic and fluent broadcaster. In Essex his paintings seemed to soften and become more painterly, but in 1951 he also extended his interests into yet another sphere when he collaborated with Basil Wright to make the memorable film on *The Drawings of Leonardi da Vinci*. Further visits to Italy, as well as to Spain (1953) and America (1954) contributed to his already wide experience, but the watershed was undoubtedly the large retrospective exhibition of his work held in the Whitechapel Art Gallery in 1955. It received a distinctly mixed reception, but most important of all it encouraged Michael Ayrton to assess dispassionately what he had achieved in the course of the previous ten years and what his future lines of investigation were to be.

During his first visit to Italy, in 1947, he had been concerned primarily with learning more about the Italian Renaissance and its background, and in the course of these studies he discovered for himself the sculpture of Giovanni Pisano. At first, the impact of Giovanni Pisano was revealed by a change in character of the imagery of Michael Ayrton's paintings, away from the linearity of the northern Gothic tradition (stemming from his interest in Dürer and Grünewald), towards the monumentality and plasticity of Italian Renaissance painting, though the development was by no means always consistent. Michael Ayrton's move towards sculpture during the early 1950s was, nevertheless, a more complex process and his admiration for Giovanni Pisano was only part of it. Ever since his stay in Vienna, drawing had been of central importance to his creative processes, and the lessons which he had learnt from his first-hand studies in the Albertina had borne fruit when he taught drawing at Camberwell. Then, and until the end of his life, Michael Ayrton made a particular point of drawing every day, and as he explored various themes through his drawings it became increasingly clear that the most effective means of expression for the ideas which immediately concerned him lay in the medium of sculpture. Indeed his drawings led him to a situation where he had to move into three dimensions or the free flow of his ideas would have been limited

89 Snowbird, 1955

to the projection of sculptural forms onto a two-dimensional surface or stifled altogether. Michael Ayrton himself stated the problem clearly in 1959, 'There are times when one's drawings, coming out from the paper, *insist* on becoming sculpture, at the next stage, rather than painting. Over this one has no control, unless one deliberately ignores the dictates of one's vision.'

The actual transition into sculpture was a difficult and protracted process, and while hesitating on the brink, Michael Ayrton fed back many sculptural ideas into his paintings. This, in turn, gave his paintings a new authority and further delayed his move into three dimensions. However, during a Manzu exhibition in London, Michael Ayrton watched a blind girl reading the sculptures with her finger-tips, and her appreciation of the forms and textures of the bronzes deeply moved him. This experience is recorded in his painting *Blind Girl with Sculpture* of 1952-53. His earliest surviving sculpture, *Snowbird,* was not made until 1954 and then only as a prop for a still life painting, never intended to have an independent existence. Inspired by Picasso's composites and certain of the forms exploited by Graham Sutherland, the bird's head is provided by a rat's skull with a body formed of a rabbit's skull. The perch is a piece of weathered wood, and it was not until two years later that Michael Ayrton had the idea of casting it in bronze. Meanwhile the painting depicting it, *Snowbird* (1954), in which the landscape elements are treated in a style continuing that of *Winter,* was acquired by the Philadelphia Museum of Art.

In a totally different vein, Michael Ayrton modelled in wax *Scavenger I* in 1954, and his emaciated dog reveals a considerable surety of handling and technical competence. However, he never received any formal training in technique and by his own admission his first attempts to model in clay were little better than children's mud pies. It was the technical advice of Henry Moore, given freely over many months, which enabled Michael Ayrton to obtain the basic, practical skills required for making sculpture, and Moore showed him how to work direct in plaster on iron armatures. Of greater importance still for the future, was Henry Moore's demonstration of the properties of the wax which he uses for certain small *maquettes,* and which can be cast direct. This technique rapidly became Michael Ayrton's preferred medium for small bronzes, and he always attached great importance to executing personally the final surface working of all his waxes prior to casting. *Scavenger I* was shown as a wax model at the Matthiesen Gallery, London, in 1961, and was partially remodelled as *Scavenger II* before casting in 1968, but it occupies a particularly important place in Michael Ayrton's development since it was the first occasion on which he deliberately set out to do a piece of sculpture totally independent of his paintings. This was in response to a challenge from Henry Moore, who came over to see it and remarked that it was 'a little bit melodramatic and like Doré'. He was, however, very encouraging and Ayrton then turned to modelling the human figure in wax in the light of the technical advice he had received.

Originally conceived as little more than an exercise, *Figure on a Wall* (1954) proved to be a highly important seminal composition for both Michael Ayrton and Henry Moore. For Michael Ayrton the male nude figure led on to an important series of figures and compositions, including *The Bathers,* while Henry Moore observed on seeing the wall, 'That's a bloody good idea; I'll use that!' The use of bones as a sculptural medium, as against a source of pictorial ideas, was also suggested by Henry Moore and provided a second line of development, though related to *Snowbird* in materials only. But whereas for Henry Moore the forms of the bones are a source of inspiration for his vitalist sculptures, Michael Ayrton always used them as a visual metaphor for landscape or for armour. *Talos Armed Head I* of 1954-56 falls into the latter class, and there is also a certain debt to the ideas of Wyndham Lewis, whom he knew best in the mid 1950s. The fierce archaism of the forms is further exploited in Michael Ayrton's monotype *Armed Head, Study for Bronze* of 1955, while *Talos Armed Head II* (1957) is much more consciously Greek in inspiration and the 'antique' surface was an experiment in the so-called 'Marini' patina which he never used again. This particular model is rather closer to Henry Moore in conception than the earlier versions, but the bones suggested the metaphor, rather than Ayrton's searching out forms in bones to express a pre-existing idea. Throughout his career as a sculptor, Michael Ayrton was fascinated by the language of bones, and the slightly sinister imagery of the *Talos Armed Heads* derives much of its power from the subtle clash between the rationalism of the Classical world and the metamorphosed animal forms. In Michael Ayrton's own words 'the bone is so directly relevant to function, so absolute in shape, that it is, in itself, transcendental sculpture'. A bone invites metamorphosis and recreates itself in the process. The skeleton of a bird can become the bare trees of a mysterious landscape and the skull of a rabbit or the breastbone of a goose will show you how it can become a helmet or wings or a fish. It will paraphrase a state of mind, arm a warrior, teach the ancient history of a species in a cryptic and ironic language'.

90 Scavenger I, 1954

91 Snowbird, 1954/56

92 Armed Head, Study for Bronze, 1955

93 Portrait of Henry Moore, 1967

94 Talos, Armed Head — I, 1954/56

95 *Stylite, 1958*

96 *Figure on a Wall, 1954*

97 *Still Life, Skull and Sculpture, 1954*

When human figures are combined with bones in a single composition the evocative ambiguities of scale can be the source of strangely powerful imagery. In *Stylite,* executed in 1958, the simple nude figure seems to grow out of the massive leg bone, at once part of it and yet separate, while in *Orpheus Seeking I* (1958) the bone is employed as a visual metaphor for a strange and mysterious landscape. Here, the pelvic girdle carries a special connotation, and the bone is identified with Gaia, the earth-mother, and hence with the very earth itself. In certain of Henry Moore's great reclining nudes he identifies the landscape with woman, and by implication moves towards the mother/earth concept, but his approach is one of a fundamentalist while Michael Ayrton's approach was conditioned by Classical cosmology. Ayrton's landscapes are generally more masculine in spirit, although there is no great significance either way in this divergence of conception. Nevertheless, Michael Ayrton for the rest of his life revitalized his ideas by renewed contacts with bone forms, and the skull of a Nile Perch which he was given in 1974 inspired *Sea Figure.* This, the skull of the largest freshwater species in the world, came from Lake Rudolf, close to the Olduvai Gorge where the earliest remains of primitive man have been discovered. To Michael Ayrton the shape of the skull suggested the visual

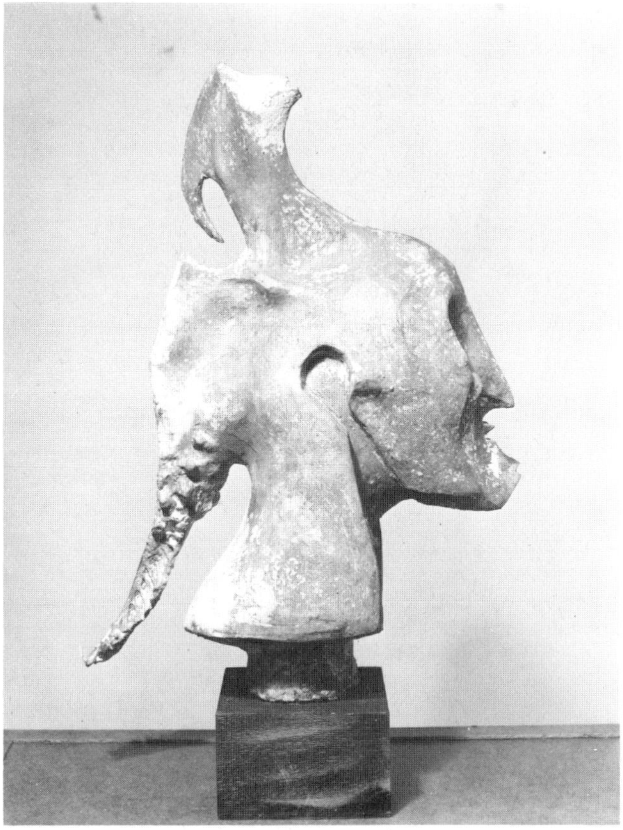

98 Talos, Armed Head—II 1957

99 Orpheus Seeking I, 1958

100 Orpheus Seeking II, 1971

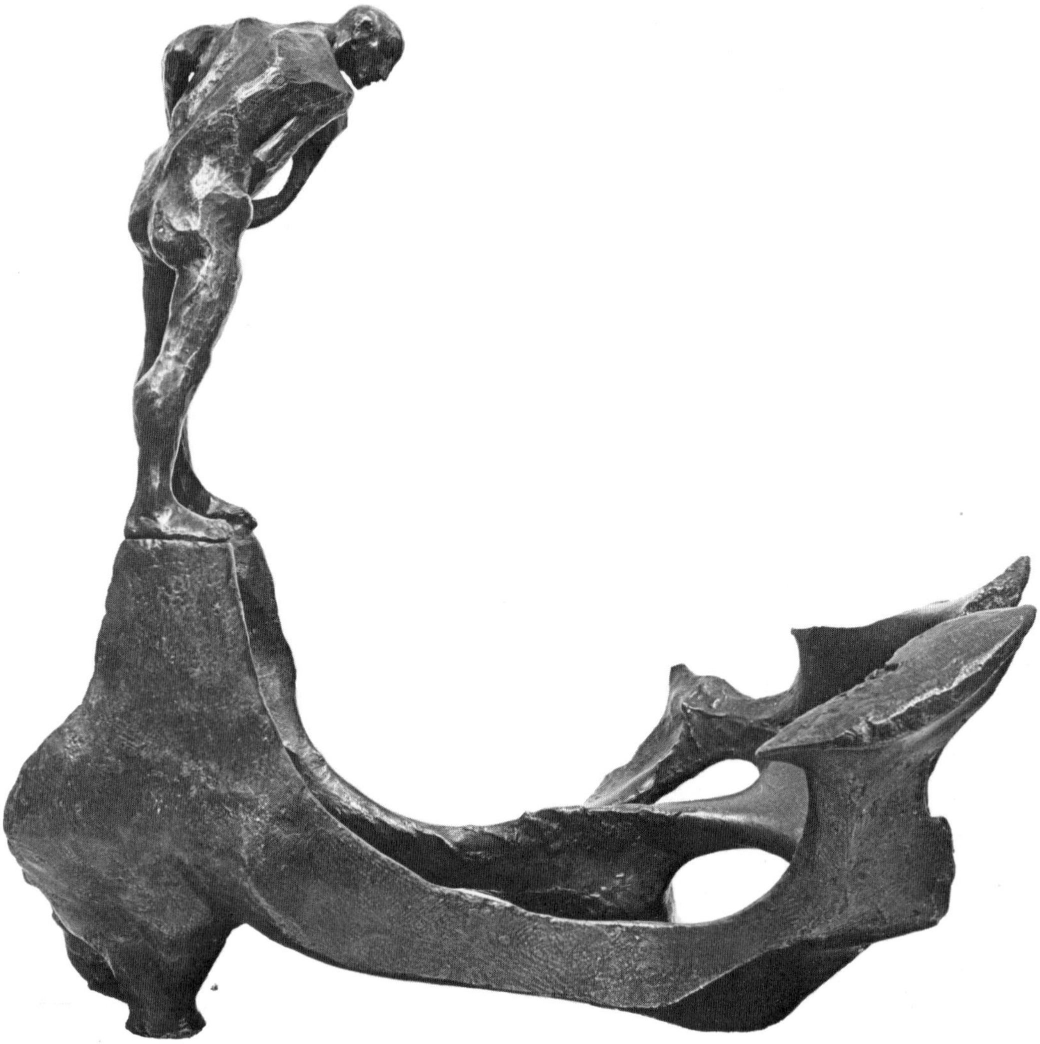

101 Sea Figure, 1974

102 *Falcon Watcher, 1954/72*

103 *Shepherd, 1954*

metaphor of man emerging out of a bone specifically associated with this area of Africa, and these ideas are explored more explicitly in a drawing, *Sea Figure,* now in the USA. In the event, the associations of the skull provided justification for an idea suggested by the bone forms, and the model also illustrates the theme of 'time' running through most of his work.

To return to the series of figures which descend from *Figure on the Wall,* Michael Ayrton's first attempt at a moderately large sculpture was the *Shepherd* executed later in the same year (1954). The theme of the lamb across the shoulders of the shepherd had been explored earlier in his painting *The Shepherds* (1951, City Art Gallery, Manchester) and in related drawings, where the treatment was more or less naturalistic, but the bronze *Shepherd* reveals more clearly the direct impact of the Classical prototype. In many respects Michael Ayrton conceived his *Shepherd* as an *hommage* to the ancient sculpture, and thus there is no Christian symbolism, conscious or unconscious, and in fact no symbolism whatsoever. Apart from the technical difficulties over which Michael Ayrton received such generous support from Henry Moore, the compositional problems to be faced by the painter-turned-sculptor were formidable and took several years to solve satisfactorily. He turned to Donatello, Degas and Rodin in particular, and his first life-size bronze figure, *Falcon Watcher* (1954), is appreciably more sophisticated than his *Shepherd.* Originally exhibited in the Whitechapel Exhibition in 1955 as a plaster under the title of *Figure watching the Flight of a Bird,* one arm was changed and other slight modifications were made before casting in 1972, so part of the figure's present appearance is due to the later reworking. No comparable difficulties of interpretation apply to his *Figures on a Jetty* executed in 1954-55 following a visit to Venice. Michael Ayrton's first multifigure composition, this deceptively simple group was conceived as an exercise in the interrelationships between figures reacting to the same set of circumstances. He had watched a group of boys and young men bathing from a jetty on the Giudecca, and prepared a series of studies of the human figure in gentle movement and at rest. The ostensible subject-matter of the group is thus of little significance, and here Michael Ayrton was interested in the human figure to the exclusion of almost everything else. Leading from these studies and the *Shepherd* is the large *Bather with Child* of 1956-57 in the collection of Mrs F. G. Miles, of which the *maquette* of 1956 is rather more successful. Despite the superficial resemblance to Henry Moore's earlier treatment of this theme, Michael Ayrton did not share the same interest in Central American sculpture and Henry Moore's great influence on the immediately post-war generation of sculptors—including Michael Ayrton—is easily misunderstood. His influence has been primarily as friend and mentor rather than as teacher to dozens of aspiring sculptors, and his personal support and encouragement crucial for the blossoming of English sculpture since the late 1940s.

Parallel to the development of the *Bathers* series of sculptures, Michael Ayrton was concerned in 1954-58 with human figures under extreme stress.

104 Bather with Child II, 1956

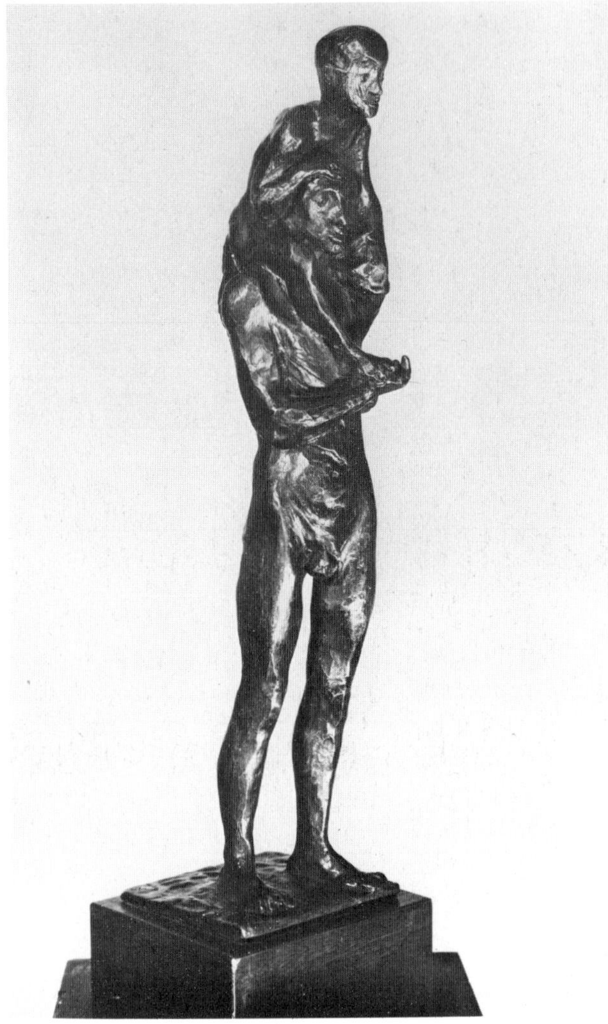

105 Bather with Child, 1956

106 Study for Bather with Child III, 1956

107 *Figures on a Jetty, 1954/55*

108 *The Swimmer, 1952*

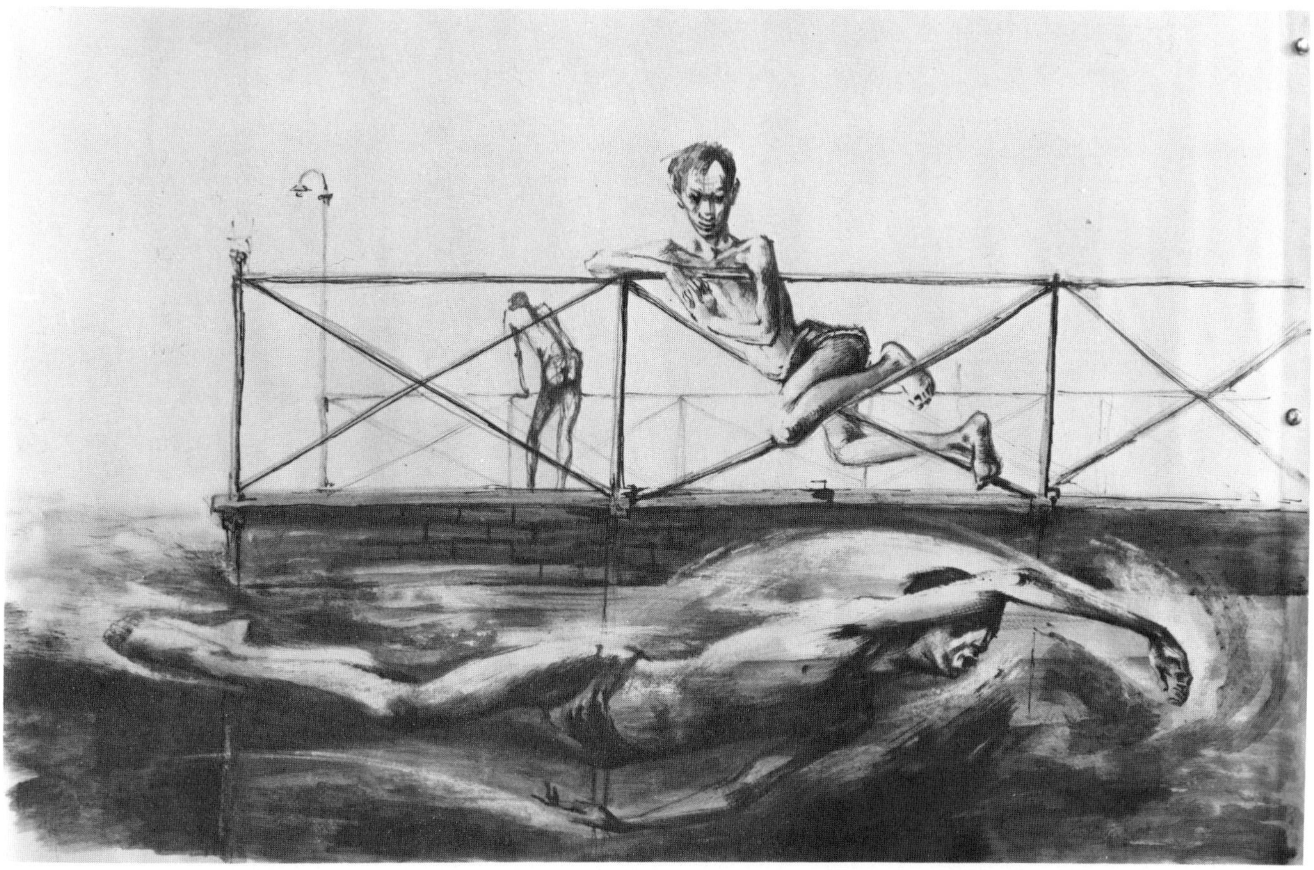

(facing page) *109 Bather with a Child, 1956* (detail)

110 Handstand, 1956/57 (detail)

The long series of bronzes and drawings depicting *Figures in Balance* draws its imagery from the world of the circus, and to this extent echoes the subject matter of his earliest paintings. Central to it, however, was Michael Ayrton's interest in the logistics of the human structure. These lithe and twisted acrobats, balancing on springy steel rods, have also a personal significance in that subsequent events have revealed them as a subconscious urge for violent movement motivated by the sculptor's growing arthritic condition which ultimately found an agonized personal expression in *Treadmill* (1974). The momentary equilibrium in the midst of tension and movement is superbly captured, and this quality is enhanced by the gently quivering motion of the bronzes if slightly disturbed. In 1953, for the first time since the Civil War, Michael Ayrton returned to Spain, but the specific influences of this experience remained relatively insignificant and the paintings developing Spanish themes mostly date from 1955-1957. Certain subject matter, such as his interest in dwarfs, can be traced back to Spain (see *Catalan Cage Birds* and related studies and prints of 1955) though paintings such as *Pipe Smoker on the Tagus* (1956) with a view of Toledo behind were most exceptional, and goats were to become a regular feature of his later landscapes. Between 1954 and 1959, Michael Ayrton achieved maturity as a sculptor and the contrast between

111 Figure in Balance III, 1956 (detail)

112 Double Scissored Figures, 1956

Figure on a Wall and the last of this series, *Breakwater,* executed in 1959, reveals the enormous distance which he travelled during those five years.

The Greek elements already noted in the *Talos Armed Heads* of 1956-57 were in part due to another extended visit to Italy when Michael Ayrton turned away from the Italian Renaissance and instead began to explore the ancient world of Southern Italy. In May 1956 he spent three days at Cumae near the Bay of Naples, and this ancient site remained from then the most seminal place as *genus loci* for Michael Ayrton. The earliest Greek colony in Italy, Cumae is inseparably associated with the Sibyl whose oracular cavern still exists below the almost impregnable citadel, and the view down the isthmus connecting Cumae to the mainland fascinated Ayrton. During those few days he produced a mass of drawings exploring this particular view of the Acropolis where for him Virgil, Berlioz and the myth of Daedalus came together in a crescendo. For Michael Ayrton 'Virgil was one of Berlioz's many gifts' to him, and he recalled:

> 'It was not Virgil's epic, it was Berlioz's setting of it that rang in my ears when I first entered the Acropolis of Cumae to find myself in thrall to a Greek myth which ends in Italy. It was at Cumae that Aeneas first landed in Italy. But it was quite a different myth that had brought me there. The myth with which I came to be concerned and which indeed inspired almost all my work for upwards of thirteen years was that of Daedalus the archetypal craftsman who built the labyrinth on Crete, to house the Minotaur, and was himself inprisoned in it with his son Icarus. In the labyrinth he made wings for himself and his son and they became the first men to fly. At the end of that flight, during which Icarus died in contest with Apollo, it was at Cumae that Daedalus landed.'

113 Acrobats II, Variant, 1958 (detail)

114 Treadmill, 1974

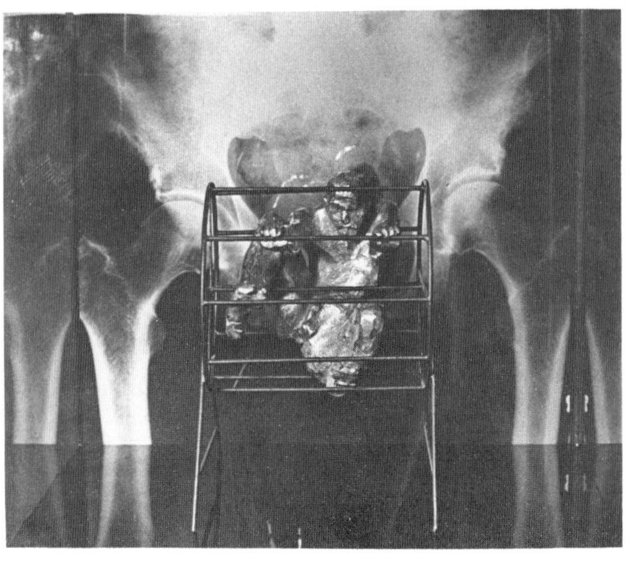

115 Parade, 1956

116 Catalan Cage Birds, 1955

Michael Ayrton's deeply rooted love for the music of Berlioz went back to his youth, and apart from *The Trojans* of Berlioz this climactic experience at Cumae drew so many other ideas together. Landscape suddenly returned again to the forefront of his interests and he continued to produce large numbers of superb landscape drawings, mostly in pen and wash, as he explored Greece and the Eastern Mediterranean in the succeeding years. The feedback into his sculpture was direct and intense, and bronzes such as *Figure in a Landscape* and *Watching Figure* (both 1957) were executed in direct response to his experience of the landscape of Southern Italy. Once again, bones are used as visual metaphors

117 Cumae, 1956

118 The Acropolis at Cumae, 1956

119 *Mycenae — the Citadel, 1957*

120 *From Delphi towards Itea, 1958*

121 Dog at Phaestos, 1957

and the reintegration of painting and sculpture was to be important for the future, not least in the genesis of his wax reliefs.

The Oxokerite, or amber-coloured 'earth wax' included in the mixture employed by bronze sculptors for the *cire perdue* process is too hard and brittle to be usable in its pure form, and beeswax, resin and turpentine are added to improve its handling qualities, though the resulting composition tends to be opaque and lacking in surface transparency. In 1958 Michael Ayrton was introduced to a highly refined form of Oxokerite, trade-named Ceresina Alb, which is white and strikingly translucent. This material adheres firmly to wooden surfaces, and by tinting it and adding resin for strength Michael Ayrton was able to obtain a material with a range of colour from white, through honey and the red-gold of Oxokerite itself, to a dense black when acetylene soot was added. The latter was obtained by shutting off the oxygen supply to the oxy-acetylene torch used as the source of heat in manipulating the wax, while the adhesive strength of the wax compositions was sufficiently great for bones and other inert materials such as broken glass, eggshell and dried sea weed to be imbedded so as to form collages.

As is so often the case, the technical innovation was evolved at a critical moment in Michael Ayrton's artistic evolution, and it immediately followed the impact of his extended stay in Greece that year. His experiences at Cumae were confirmed by the six months he spent in Greece as an adviser to Basil Wright in the making of the film *The Immortal Land,* and a by-product of this project was the prize-winning film *Greek Sculpture* which they co-directed. As Michael Ayrton himself was first to admit, 'To make a film is a lengthy and laborious business. To film sculpture necessitates looking at it harder and for longer than almost any other discipline', and 'It was in making this film *Greek Sculpture* that Ayrton learned to know the subject'. In *Figure in a Landscape* his revived interest in landscape *per se,* following the visit to Cumae, was expressed in the new relationship between the figure and his environment, but in Michael Ayrton's twelve wax reliefs of 1958-59 the synthesis of his painting and sculpture into a single medium took a new form.

In *Sphinx* (1958) much of the imagery of the *Talos Armed Heads* is further refined and the sharp sutures of the bones contrast with the soft flowing patterns of the wax. Out of this distant and mysterious landscape the sphinx seems to have condensed, at once the product of it and its mistress. In the earlier *Talos Armed Heads,* the angular forms and dry surfaces epitomized the archaic lifeless-

122 Figure in Landscape, 1958

123 Perilous Place, 1958

ness of the former guardian of Crete, but in the *Sphinx* relief the delicate nuances of form and colour, and the soft surfaces, are instead redolent of the mysterious and ambiguous utterances of that mythical being. In *Perilous Place* (1958) full reign is given to the sinister and the macabre. The spiky collage of bones, half-buried in the wax and obscured as if by drifting Northern mists, conjures up a vision of primaeval desolation. In it there is no hint of the Classical world, but instead the bone forms on the right take up the outline of a grotesque and ghostly crucified figure. Michael Ayrton's earliest allegiance was to the Late Gothic world of the early sixteenth century German painters and the *Perilous Place* provides a complete antithesis to the *Sphinx*. In both, the wax is modelled in layers using

124 Landscape of Cain, 1958

125 Midsummer, 1958

126 The Birth of Phanes II, 1959

127 The Birth of Phanes, 1959

an oxy-acetylene torch so that wax darkened with acetylene soot is overlaid with translucent layers of white wax and dribbled wax in both colours are intermingled. The velvety blackness of the acetylene soot has the added advantage of creating deep spatial effects so that the *Sphinx* is isolated from the depths of her lair, as is the centre figure in *Landscape of Cain* (1958).

By contrast *Midsummer* (1958) with a collage of breast bones and the carapace of a crab is sun-drenched and optimistic. In his wax and bone reliefs Michael Ayrton was extremely sensitive to mood and one of his great strengths was that he could be both passionately sensual and coolly detached, and the pendant to *Midsummer* is the ruin and desolation of *Midwinter* (1959) under the shadow of a dark cloud. His growing authority in this new medium is revealed well by his two reliefs devoted to the *Birth of Phanes,* and both include broken glass and a bird's egg as constituents of the collage. According to the Orphic cosmogony Chronos, the Time which never grows old, formed an egg in the Aether and from this sprang Phanes, the creator and first king of the gods. Aristophanes, in the *Birds,* expounds a slightly different variant, and states that black-winged Night bore a wind-egg in the bosom of Darkness and from this emerged Eros. Both versions of the Orphic theme are embodied in Michael Ayrton's imagery, and while the vigorous black bird forms of *Birth of Phanes I* (1958) are reinforced by a bird's skull and tendrils of dark sea weed, in *Birth of Phanes II* (1959) these forms become diaphanous and all the energy of the composition is focussed on the egg out of which Phanes was born.

Most of Michael Ayrton's work on the wax reliefs was undertaken in a burst of activity around Christmas 1958 so that the *Birth of Phanes* reliefs are in effect almost coeval, and in April 1959 he explored further the theme with a collage, of cut paper and broken glass combined with pen and ink, of Phanes emerging from the egg. This was a period of considerable experimentation and apart from the small group of white collages using Plaster of Paris instead of wax, he evolved an entirely new way of using paper collage. In *Icarus Falls* (1959), the first image he created in a new series, he used translucent coloured tissues to gain similar effects to the translucent wax reliefs and combined these with the more orthodox painting techniques. Thus the tissue was being used more like a painter's glaze than any sculptural technique, and the blackness was intensified by a whiff of acetylene smoke. The latter was fortuitous, although it serves to emphasize the extent to which the traditional boundaries of techniques and disciplines had been broken down. The paper collage leads, however, to the large scale wax relief, *Icarus Falls* (1959) fomerly in the collection of Gavin Maxwell, which provides the climax of the series of wax reliefs and which fortunately survived the fire which destroyed Sandaig in 1967. Stripped of his wax and feathers, and his wings reduced to bony frames, Icarus plummets earthwards, the victim of his own folly. The distant sun is partially obscured by the smouldering trail of the figure falling through the bright sky and the very choice of wax as the principal medium creates an additional complex visual metaphor. The transparency of the wax conveys the brilliant light of the Mediterranean, but in the last relief of the series *Icarus Drowned* (1960), in the collection of the National Gallery of New South Wales, with the dead Icarus amid the wreckage of his wings, modelled Plaster of Paris was chosen instead. Failure and death, the mood is one of intense pessimism.

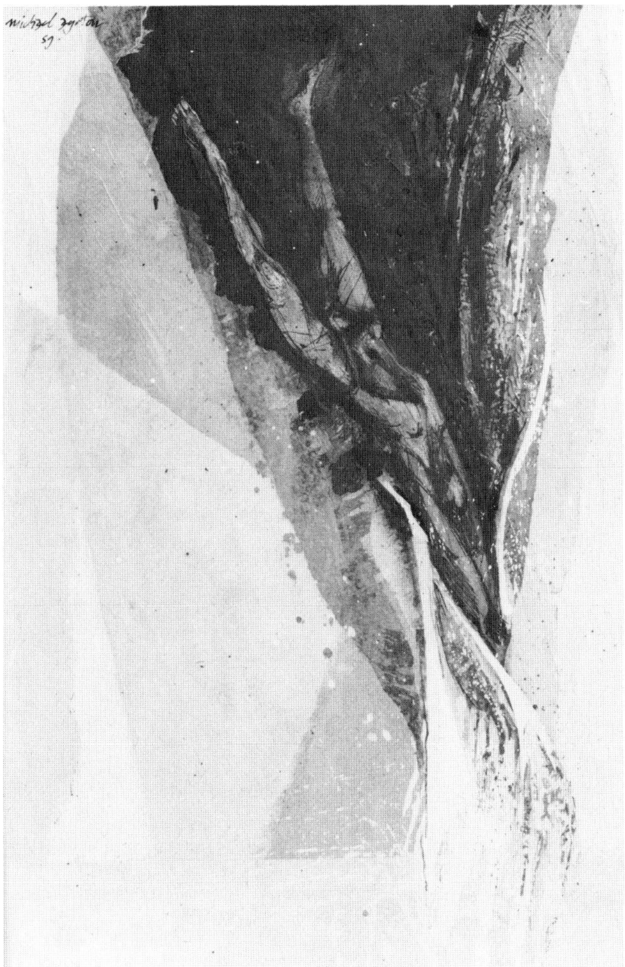

128 Icarus Falls, 1959

The development of Michael Ayrton's painting style after the Spanish sequence closely reflects the evolution of new techniques for the wax reliefs and collages, but during the winter of 1958-59 he painted several important works which continued his earlier style. *Slow Dance for the Nativity* is the most distinguished of these and was executed in Essex between December 1957 and February 1958. In many respects it is the last large painting executed by Michael Ayrton in the figure style he had evolved in Ischia, and the strange *Orpheus* executed in wax and broken glass over ink and wash on panel (1959) is in effect a lateral extension of the wax reliefs although the conception of the figure is relatively close to the earlier paintings. Neither is a preparation for the rich colours, broad brushwork and painterly

(left) *129 Icarus Falls, 1959* (above) *130 Icarus Drowned, 1960*

qualities of *The Terrace at Night* painted in Provence in summer 1959, and much of the special freshness and spontaneity of that painting may be attributed to its execution in oils on the spot in France and not months later in England. In strong contrast the *Icarus Falls* sequence, also painted in 1959, is executed in acrylic with extensive use of dribbled paint and spatter techniques to build up images much closer in effect to the waxes and collages. Subsequently, these techniques were used to devastating effect in the large Greek landscapes of 1959-60, such as *Large Shore* and *Large Cycladic I,* where the detail is burnt away by the intensity of the light. Textural effects were enhanced by the inclusion of sand in the acrylic medium and the dividing line between these paintings and his painted collages like *Black Shore* (1961) to all intents disappeared, while the rich textural effects were also exploited in the monotype *Dark Sea* of 1962. The Greek landscapes of 1959-62 are amongst the last paintings Michael Ayrton executed in direct response to the external appearances of the natural world, but in *Large Shore* and *Large Cycladic I* he created images of rare authority and power.

Michael Ayrton's first contact with the legend of Daedalus and Icarus had been on his visit to Cumae in 1956 and he turned to the theme of Icarus at the beginning of 1959. He soon moved away, however, from the limitations of the reliefs and collages, and with *Icarus I* (1959) deliberately moved into the exploration of the theme in terms of three-dimensional sculpture. In *Icarus I* there is a direct connection between the bone forms in the wax reliefs and the open ribbed structure of the outstretched wings, and this figure was almost coeval with the later wax reliefs. The heavier and more solemn figure of *Daedalus Winged* (1959-60) followed soon afterwards and the much larger figure of *Icarus III* (1960) and

131 Slow Dance for the Nativity, 1958

Icarus III Variant also adopt a more stocky and muscular pattern. The latter was cast in 1972, with small but important variations, for presentation to the City of London by J. B. Sunley, and has been installed in Old Change Court, near St. Paul's Cathedral. Within the Daedalus/Icarus legend the role of Icarus is very limited, and in his novel *The Maze Maker* (1967) Michael Ayrton has Daedalus state, 'Icarus was first and last a fool. He did not become a man, yet he became an immortal... and when Icarus flew at the god to cast him down... the god spurned him as you would expect... I loved Icarus and I have tried to understand him... It was

132 Large Shore, 1959/60

(above) *133 The Terrace at Night, 1959* (right) *134 Icarus Falls III, 1959*

essential to him to become a hero. Yet I, who bring order, measuring to make harmony, brought him to his disorder so that he fell. His flight and his wings were of my making. His flight was mine.'

In his interpretation of the myth Michael Ayrton was unable to accept the seemingly meaningless primitive account of the flight from Crete in that Icarus simply flew too close to the sun so that the wax of his wings melted and he was drowned in the Aegean Sea. Instead, Ayrton preferred to see the action of Icarus as a deliberate sexual attack on Apollo, with his inevitable defeat. These ideas are developed in a series of works executed in 1960-62, and in the painted Plaster of Paris relief *Icarus in Flight* (1960) the pride and self-confidence of the flying figure is well expressed. In his novel *Testament of Daedalus* (1962), the traditional wax and feather wings are rejected in favour of the more mundane wood, gut and sheep bladders, and elaborated further in the accounts in *The Maze Maker,* so that all but the

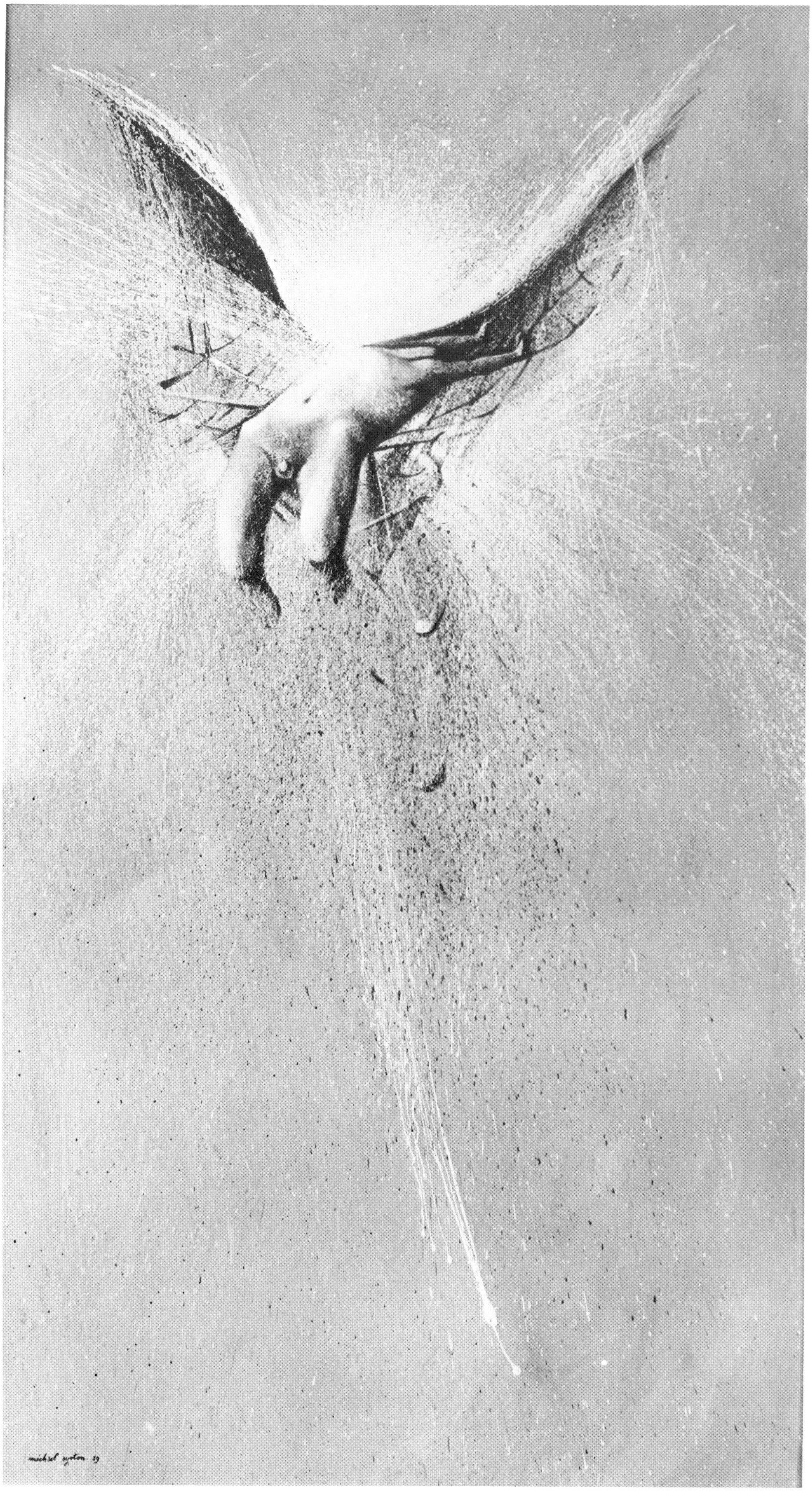

135 Large Cycladic 1, 1959/60

136 Corn Rhyton, 1961

earliest wing structures reflect the forms natural to such materials. *Icarus in Flight,* with its pattern of rearranged anatomy set against the stretched-membrane wing forms is one of the most abstract of the series and points the way to the charcoal drawing *Icarus Pinwheel* (1960) in which the figure spins helplessly on impact with the Sun. These ideas are developed further in *Icarus Transformed I* (1961) where only the massive chest and shoulder muscles which carried Icarus up to the Sun survive and the remainder of his body and wings have shrivelled up under the intense heat of the Sun's rays. The bottom half of the figure uses a fragment of bone as a visual metaphor, and the final disintegration of his figure after impact with the Sun is revealed by *Icarus Transformed III* which is modelled on part of a horse's skull—in itself a symbol of death. In 1963-65 Michael Ayrton returned to the flight of Icarus with the series of bronzes *Icarus Ascendant* and *Icarus Rising,* but his centre of interest had already moved from Icarus to the much deeper significance of Daedalus.

137 Icarus I, 1959

138 Daedalus Winged, 1959/60
(below) *139 Icarus in Flight, 1960* (right) *140 Icarus III Variant I, 1960/72*

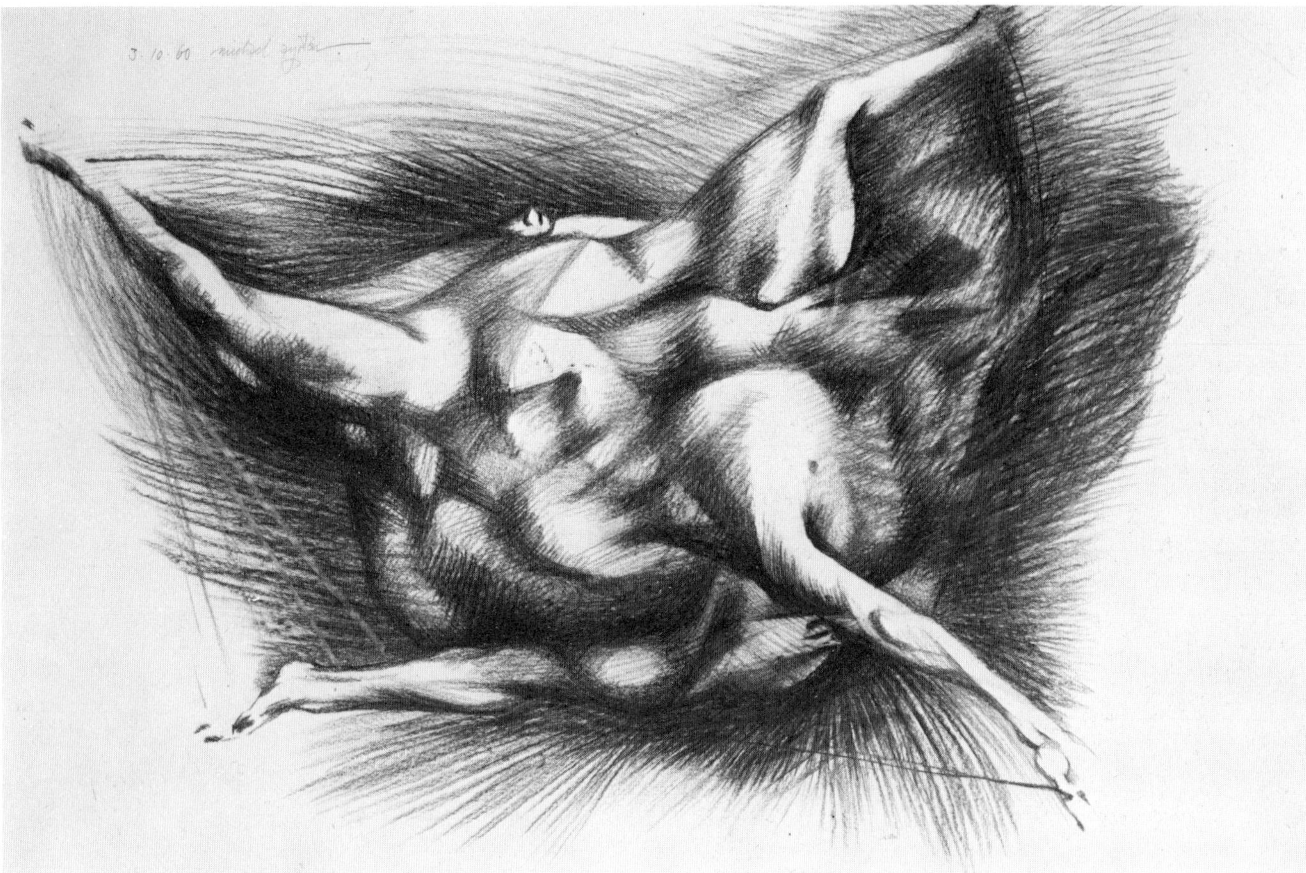

141 Icarus Pinwheel, 1960

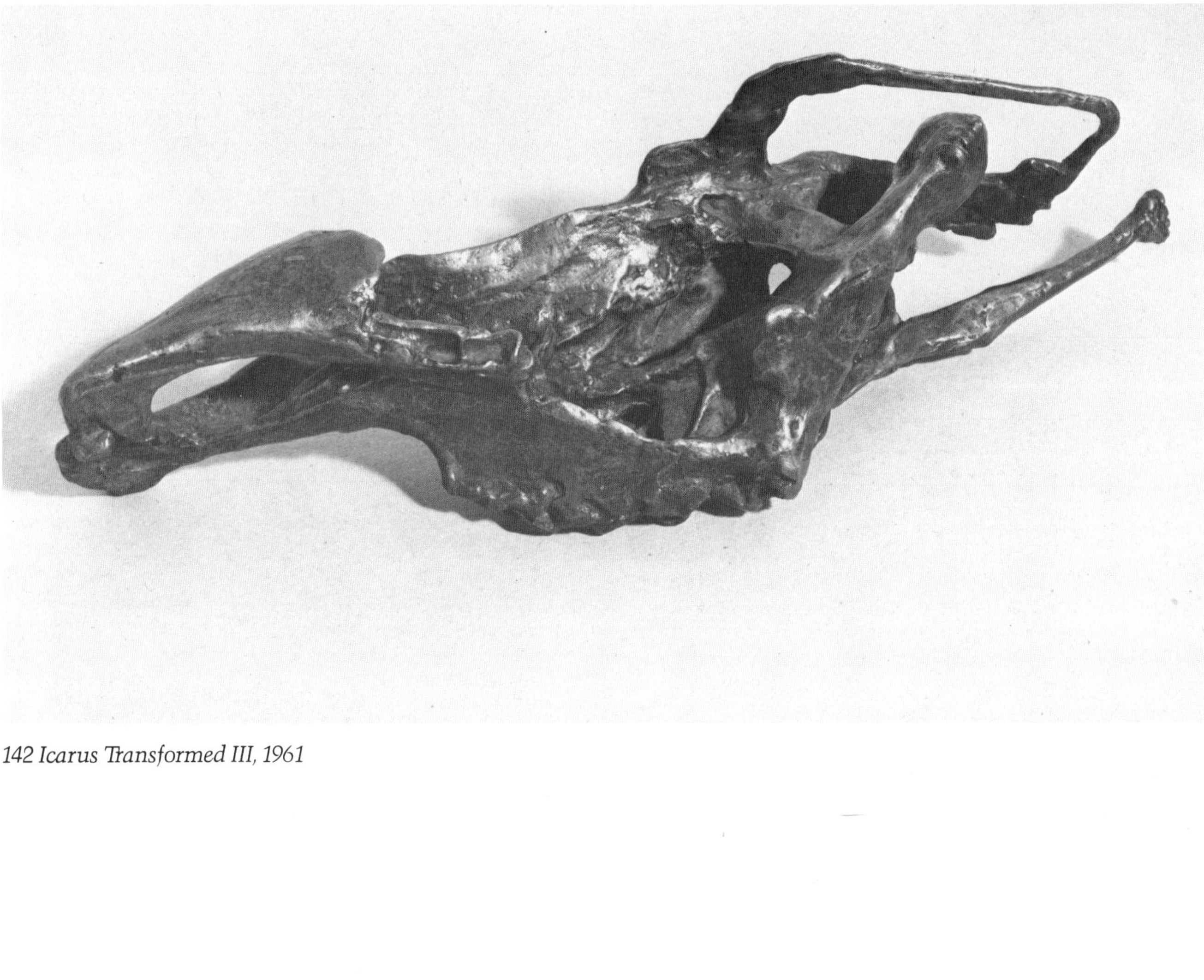

142 Icarus Transformed III, 1961

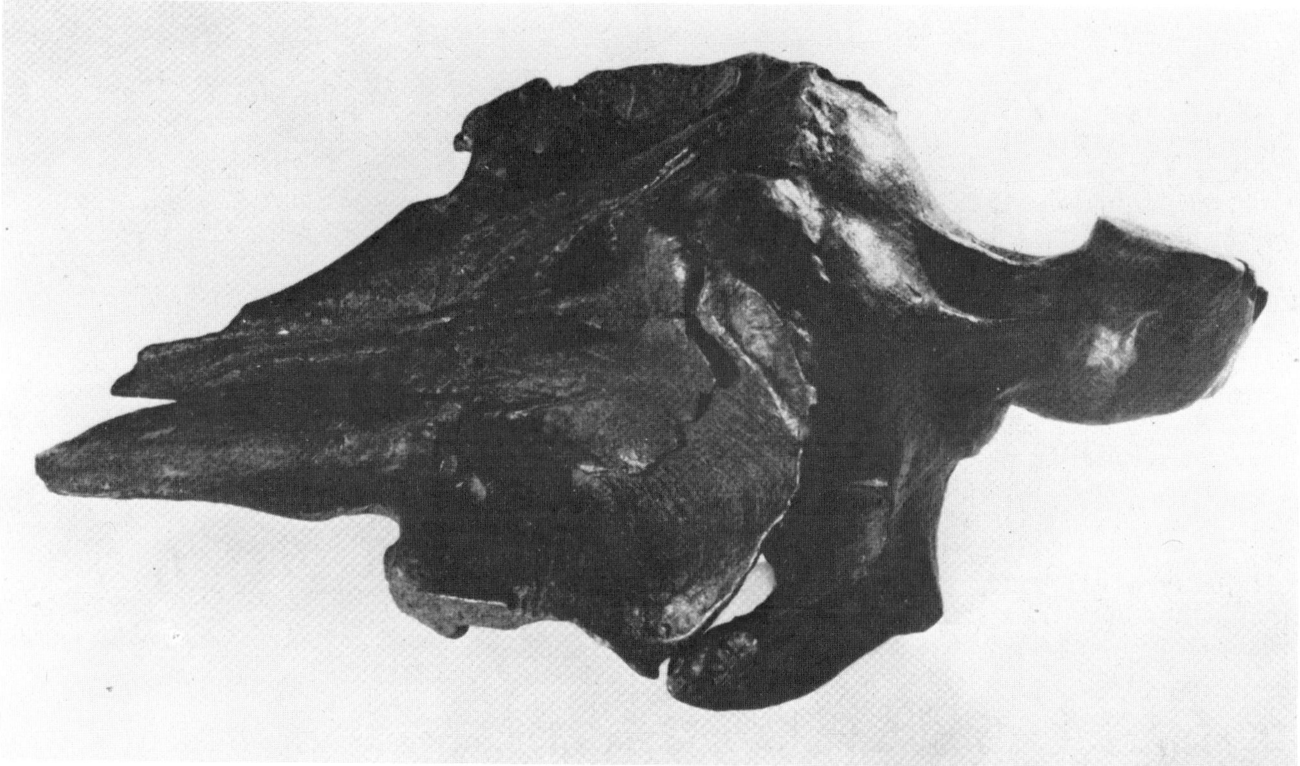

143 *Icarus Transformed I*, 1961

144 *Icarus Winged*, 1960

The word 'labrys', meaning a double-headed axe, from which the word 'labyrinthos' or labyrinth derives, is pre-Greek and labyrinth means literally the 'Place of the Double-Headed Axe'. This has always been associated with Knossos, and the discovery of simple representations of this weapon on the walls of the ruins of the palace of King Minos there tends to confirm the identification. According to the legend, Daedalus built the Maze, or labyrinth, for King Minos in order to confine the Minotaur, which had been conceived by his wife Pasiphaë, after impregnation by a bull from the sea sent by Poseidon, also assisted by Daedalus. Minos imprisoned Daedalus, together with Icarus, for his role in the latter affair, and after their escape and the death of Icarus, Daedalus flew on alone to Cumae where he built a temple to Apollo. Daedalus, however, has a greater significance as the archetypal master craftsman and inventor and this is the fundamental theme in both 'The Testament of Daedalus' and 'The Maze Maker'. The concepts of the Maze, its maker and its occupants, were to obsess Michael Ayrton for over ten years, and all the elements of 'The Place of the Double-Headed Axe' are wound together in the strange, brooding *Blade Maze Figure* of 1965.

Fundamental to Michael Ayrton's treatment of the Maze theme is the beginning of what he conceived as the 'Emerging situation': the analysis of the reactions of man to the environment of the maze, the stress of confinement and the search for release. In the studies of acrobats he had been primarily concerned with figures under extreme physical stress, accepted voluntarily by the participants as a conscious means to an end, but from the early 1960s he turned his attention to figures under mental stress and their physical impotence to relieve it. His charcoal *Figure under Stress* of 1961 anticipates many of the essential qualities of the powerful *Turning Torso,* executed in charcoal over tissue collage in 1965. In *Black Maze Figure* (1965) the figure suffers physical dislocation in response to the Maze, while in the wash drawing *Emerging Maze Figure* (1965) the reaction to the Maze is more complex. The figure is at the beginning of the way out, but a 'maze' is nevertheless still within him, and the identification of the complexity of the intestines and the convolutions of the brain with the structure of the Maze is an important departure. In 'The Mazemaker,' Daedalus identified the labyrinthine tunnels under the Acropolis at Cumae with the entrails of mother earth and closes the book with the statement 'I, Daedalus, maze maker, shall take this that I have written with me to Sardinia and dedicate it at the entrance to the maze which leads to death. Then you, before you follow me down into Gaia, who is the Mother, will know what is to be known of my journey and the fate of my son, Icarus. Before you follow me, look into the sky-maze and acknowledge Apollo who is the god.' Michael Ayrton's *Translucent Maze* (1965), executed in pen and ink and charcoal over tissue collage expresses clearly the idea of man making a maze out of himself and round himself, and it is further developed in the series of *Maze Maker* bronzes (all 1965) and the extremely powerful painting *Maze Maker II (Copper Seam)* of 1966.

The earliest bronzes depicting Daedalus, including *Daedalus the Wingmaker* (1960) and *Daedalus at*

145 *Figure under Stress, 1961*

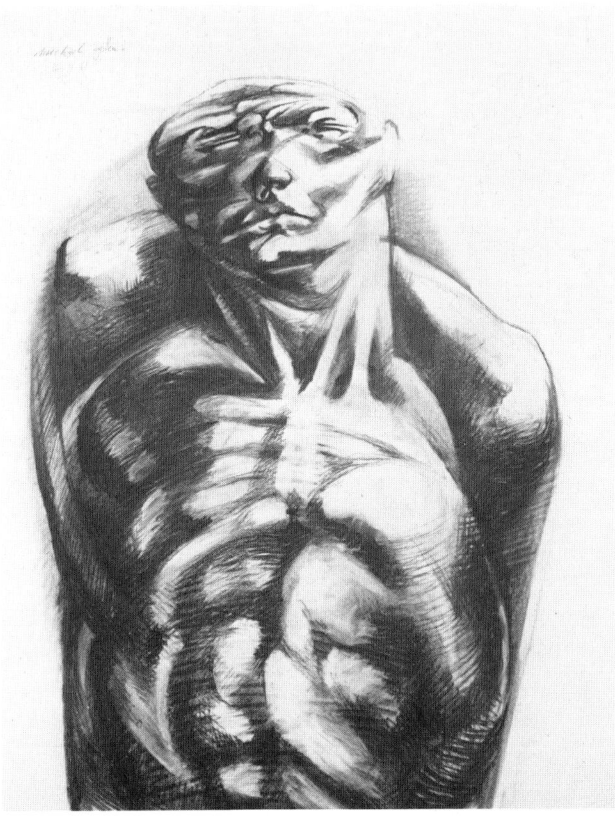

146 *Blade Maze Figure, 1965*

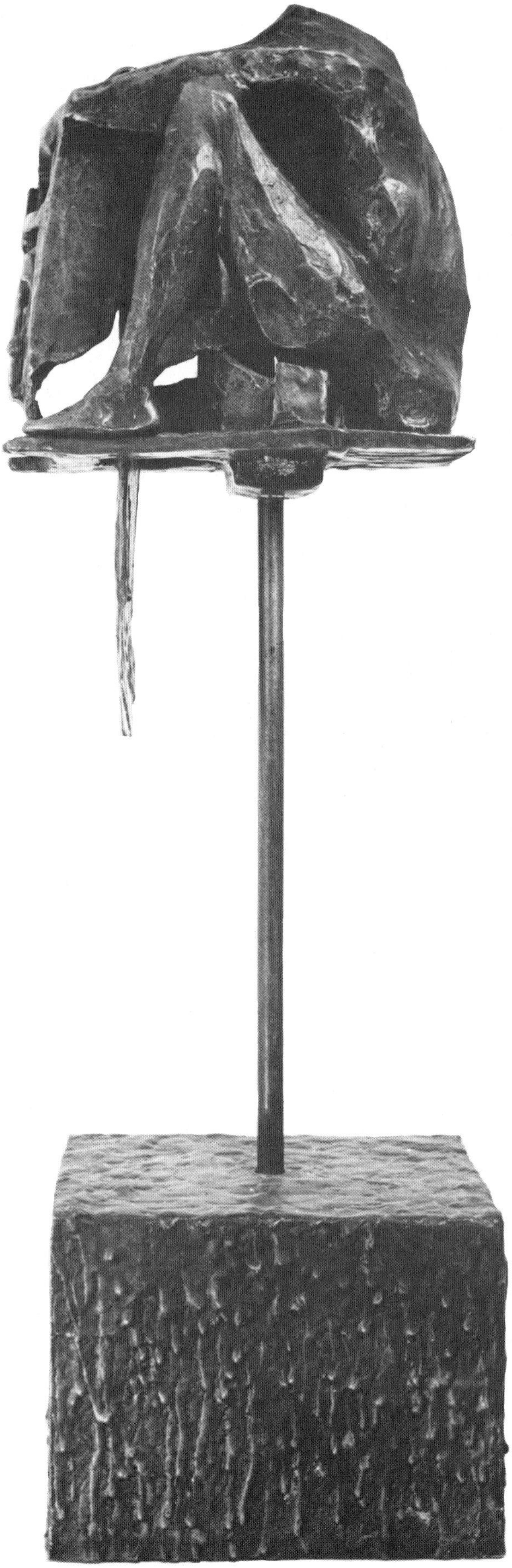

147 Black Maze Figure, 1965

Cumae (1961) are relatively uncomplicated, and the total absorption of the former is replaced by the abject misery and exhaustion of Daedalus, seated with eyes closed and empty hands, after the flight from Crete and the loss of Icarus. The image of Daedalus as the archetypal master craftsman is briefly displaced by that of the grieving human father, and the small-scale *Sun Maze* (1970) returns to the flight of Icarus from the Maze in terms of his resurrection. This composition has never been executed on the large scale which its dramatic qualities could sustain, and the main line of exploration of the relationship between Daedalus and Icarus was through the *Daedalus/Icarus Matrices.* Eventually leading to the composition for the Arkville Maze, the conception of *Daedalus/Icarus, Matrix I* (1968) is based on the idea of Icarus rising from the work of Daedalus. The image became more complex as it evolved, and although the umbilical cord is only hinted at here, it becomes

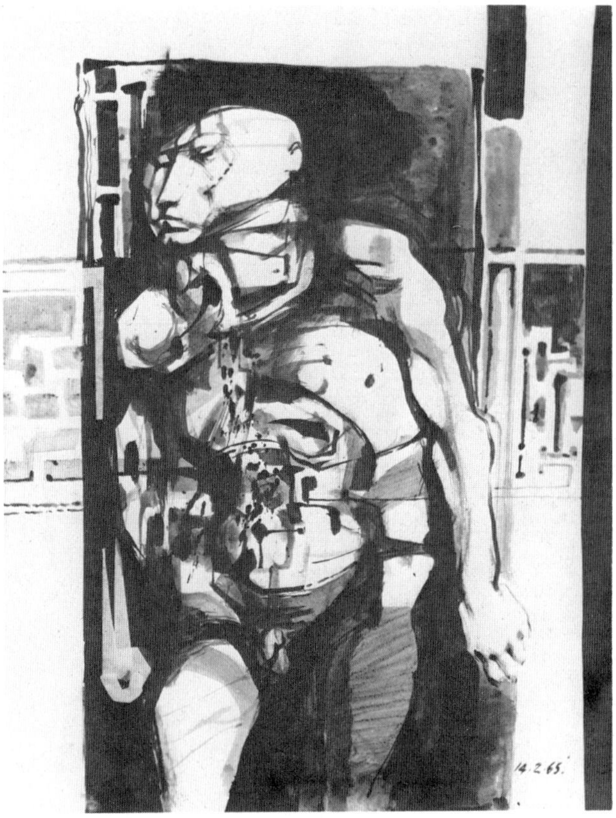

148 *Emerging Maze Figure*, 1965

a major element in *Mirror Coil* (1962) and the soaring *Daedalus/Icarus, Matrix II* (1968), as in *Icarus Rising, Variant III* (1965). In the largest version of *Daedalus/Icarus, Matrix II*, executed for the Arkville Maze in 1969, the workshop of Daedalus is more comprehensively furnished with tools and the umbilical cord is again a major feature. Michael Ayrton's exploitation of the theme of the umbilical cord in this context is based on the distinctly ambiguous aspects of Ariadne's role in the myth. The daughter of Minos and Pasiphaë, and thus a half-sister of the Minotaur, she fell in love with Theseus and used her knowledge of the maze to betray the Minotaur to him. To achieve this, Ariadne made a red thread for Theseus to find his way to the centre of the Maze and out again. This red thread is interpreted by Michael Ayrton as representing the umbilical cord, the way into the maze/womb/entrails and the way out. Indeed, he uses it as a metaphor for blood relationships between Daedalus and Icarus, as well as between Ariadne and the Minotaur, and the meaning of the ecstatic *Figure with a Skein II* (1963-72) is self-evident.

149 *Translucent Maze*, 1965

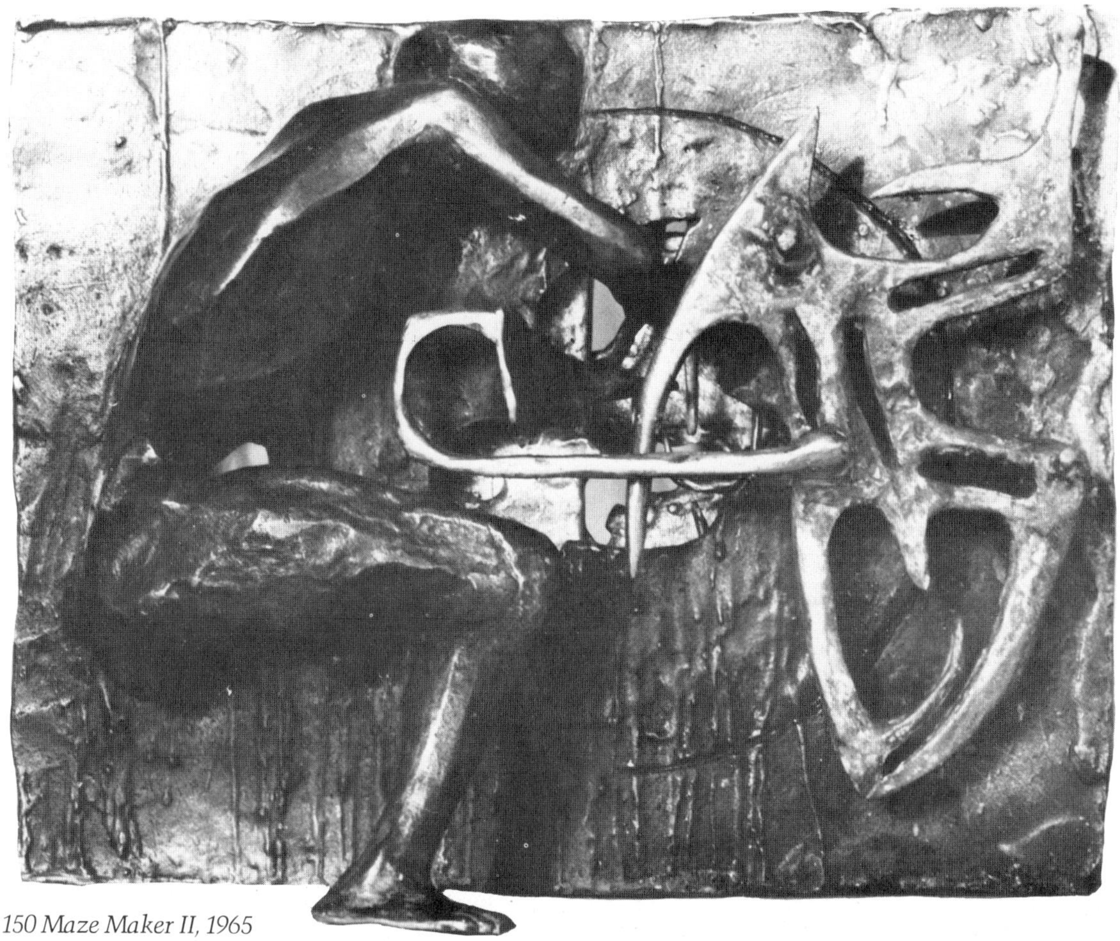

150 *Maze Maker II, 1965*

151 *Maze Maker, Large Version, 1965*

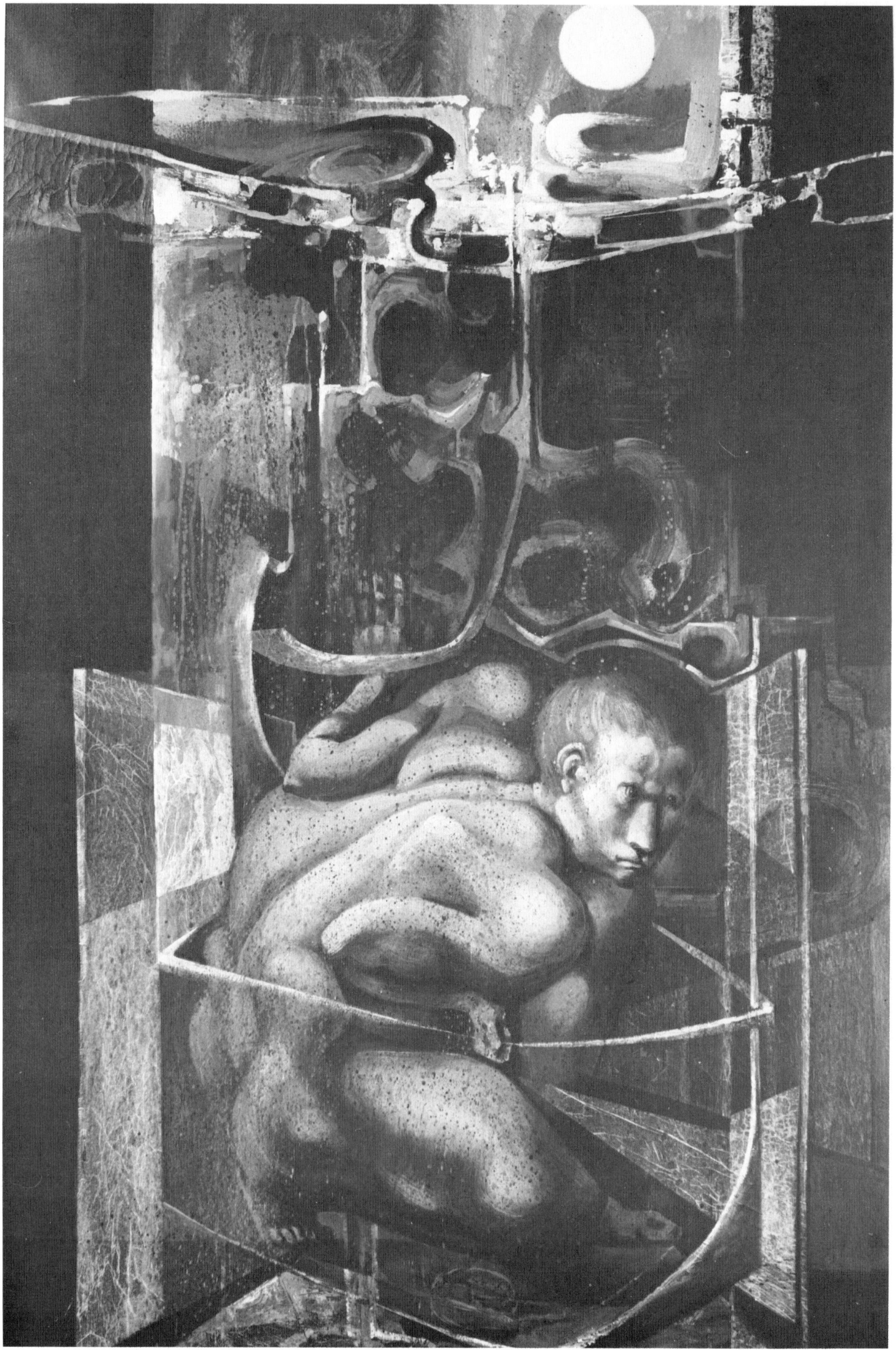

153 *Daedalus, the Wing Maker, 1960*

(left) 152 *Maze Maker II (Copper Seam), 1966*

154 *Daedalus at Cumae, 1961*

155 *Sun Maze, 1970*

head forms are more gorilla than bull, and these are motivated by pure anger and resentment. Dating from the end of 1963 and the first months of 1964, these reach a climax in the *Evolution of the Minotaur,* (1964; formerly in the collection of Lord Beaumont of Whitley), but they led to a dead end in the development of the Minotaurs and Michael Ayrton returned soon afterwards to bull-headed forms. With the third group of Minotaurs the human elements became increasingly pronounced and from now onwards human hands and feet replaced the animal hoofs. The superb series of small bronzes such as the *Restless Minotaur* and the *Minotaur in Jeopardy* (both 1964), and related drawings, provide a prelude to the great *Arkville Minotaur* (1968-69) which is the climax of the theme.

160 The Evolution of the Minotaur, 1964 (detail)

Begun in 1968, the Arkville Maze was commissioned by the late Armand G. Erpf of New York to the designs of Michael Ayrton, and its 1,680 feet of stone pathways and brick walls eight feet high were completed in 1969. Two hundred feet in diameter, it is claimed to be the largest labyrinth built since antiquity and has two central chambers. One is lined with reddish stucco, and contains the monumental *Arkville Minotaur;* the other, lined with polished bronze sheets, the largest version of the *Daedalus/ Icarus Matrix II.* Apart from the sheer scale of the enterprise, the importance of the commission lay in bringing together these two major sculptural themes into a single large scale complex and the inclusion of mirrors to fragment the images was to have great significance for the future. After 1968-69, Michael Ayrton maintained his interest in the Minotaur with a series of powerful drawings of Minotaurs under stress and in 1971 he published a set of ten etchings on the Minotaur theme, 'Asterion was thus born, consecrated, mazed and isolated. Rising, he found himself part man, pent up. Revealed he saw himself alone, part bull. Fallen, in darkness, he yet endures.' The ambiguous aspects of the myth are clearly brought out in the etchings, through the process of self-discovery of the Minotaur, and the scene of his seeing his reflection in the polished double-axe held by Ariadne is totally unprecedented (though fully in accordance with Michael Ayrton's personal interpretation of the myth). Etching was an unfamiliar medium to Michael Ayrton because he had not used it since the 1930s, but he re-learnt the techniques for this suite and the plates were printed in London by Dorothea Wight. During the last years of his life, Michael Ayrton continued to explore the theme with further drawings and small bronzes, culminating in *Minotaur Erect,* modelled and cast in 1975. The Minotaur thus remained, in his frustration at being unable to communicate fully with the world, a potent symbol for Michael Ayrton, and his deep, if not obsessive, attachment to the theme was stated unequivocably by his *Personal Janus* of 1970.

161 Restless Minotaur, 1964

162 Minotaur in Jeopardy, 1964

163 Minotaur Enraged, 1963

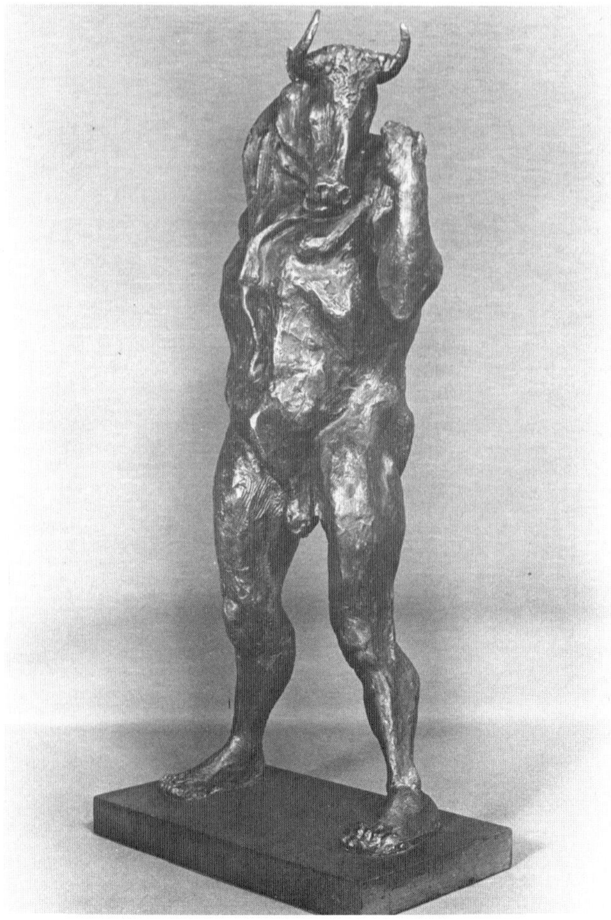

164 Minotaur Erect, 1975

165 Daedalus/Icarus Matrix Large Version, 1968/9

166 Arkville Minotaur, 1968/9

167 Minotaur Surprised, 1975

168 Minotaur Alarmed, 1963

169 Minotaur Waking

170 Minotaur Risen, 1971

171 Arkville Minotaur, 1968/69 (detail)

172 *Arkville Maze*
173 *Arkville Maze — Maquette, 1968*

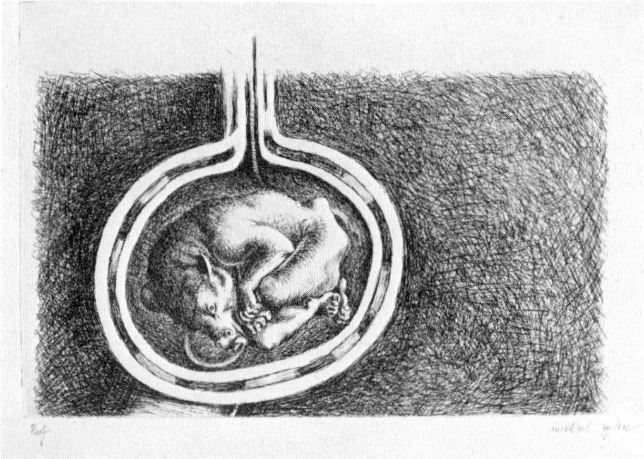

Embryo

Consecrated

As Calf

As Yearling

Rising

Risen

174 Suite of Ten Minotaur Etchings, 1971

Full Grown

Alone

Revealed

Pent

An independent development from Michael Ayrton's visit to Cumae in 1956 and his subsequent visits to the Greek sanctuaries was his series of oracles and related figures. The mysterious, sightless figure of *Oracle, Version I,* modelled and cast in bronze in 1962, is relatively straightforward as Michael Ayrton began a formal exploration of the archaic seated figures. In another model of 1962, the *Tripod Oracle,* the tripod on which the Pythia—the medium of Apollo and his oracles at Delphi—sat becomes amalgamated with the figure, and the emaciated character of the body recalls the story told by Ovid that the Cumaean Sibyl requested of Apollo as many years life as she held grains of sand. Unfortunately, she omitted to ask for perpetual youth and it is these ideas which are carried over into the bronze figure *Siren* (1963), half woman and half bird. The *Pythia* herself (1963), with tripod support, develops from *Oracle, Version I* and is the subject of a considerable group of drawings of ecstatic heads, while this element reaches its climax in the declamatory *Mantic Figure.* Within the same group of small bronzes Michael Ayrton also re-examined the myths of Talos. A nephew of Daedalus, and supposed to have been murdered by him, Talos re-emerges on coins and vases as the Guardian of Crete and the legends are explored in *The Maze Maker.* Although the large bronze figure of *Talos* (1962) was not set up in Cambridge until 1975, it and the related smaller bronzes, including *Slender Sentinel* (1963) belong to the same group as the *Oracles.* Also to be included with the Oracle group are the powerful drawings of *Demeter and Kore* (1964, 1965) and the *Seasons of Demeter* (1965) in which the relative fullness of the figure expresses the changing seasons. The ideas of the cyclical processes of nature implicit in the Demeter and Kore/Persephone myth are further explored in the massive bronze *Demeter Pregnant* (1966) and the strangely moving *Winter Demeter* (1967). The latter, partly a bone collage, refers back to some of Michael Ayrton's earliest sculptures and in accordance with the myth she is the skeletal form of winter awaiting the ritual rebirth of Persephone with the Spring. The differences from the main group of Oracles are thematic rather than formal and the Demeter sculptures and drawings were seen by Michael Ayrton as belonging with them.

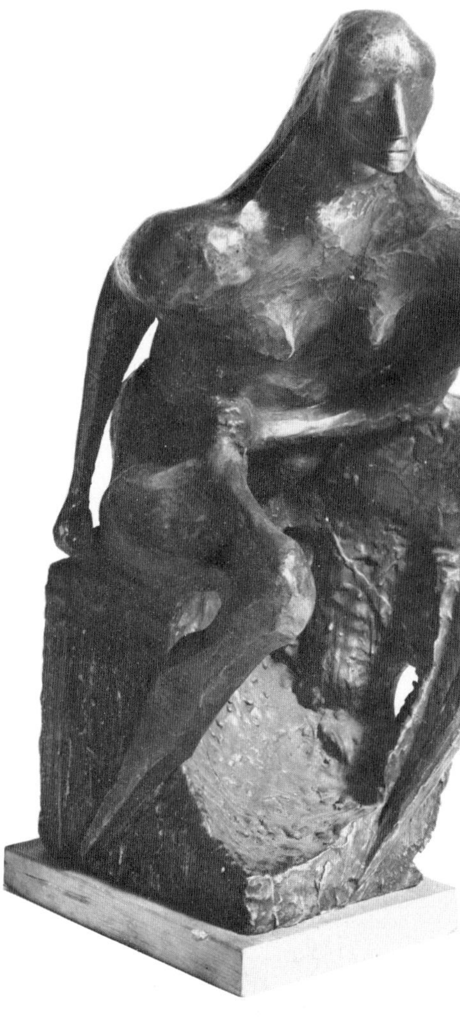

175 Oracle Version I, 1962

176 Pythia, 1963

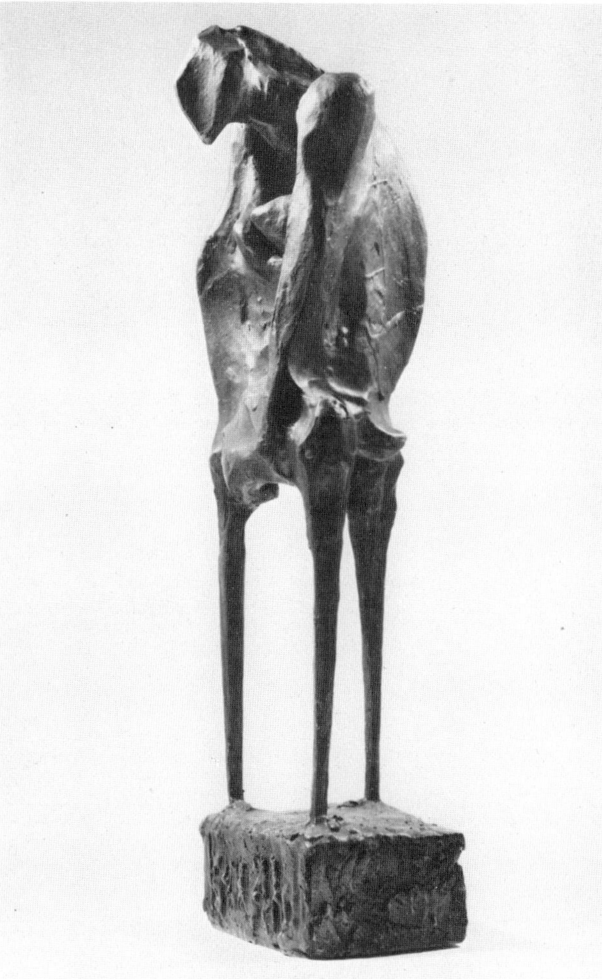

177 Tripod Oracle, 1962
179 Study for Tripod Oracle, 1962

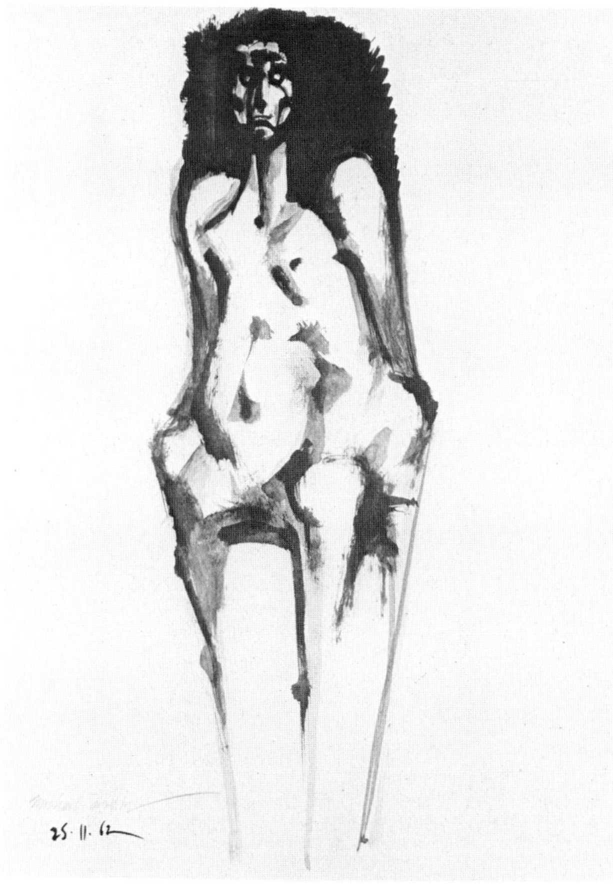

178 Mantic Figure, 1963
180 Siren, 1963

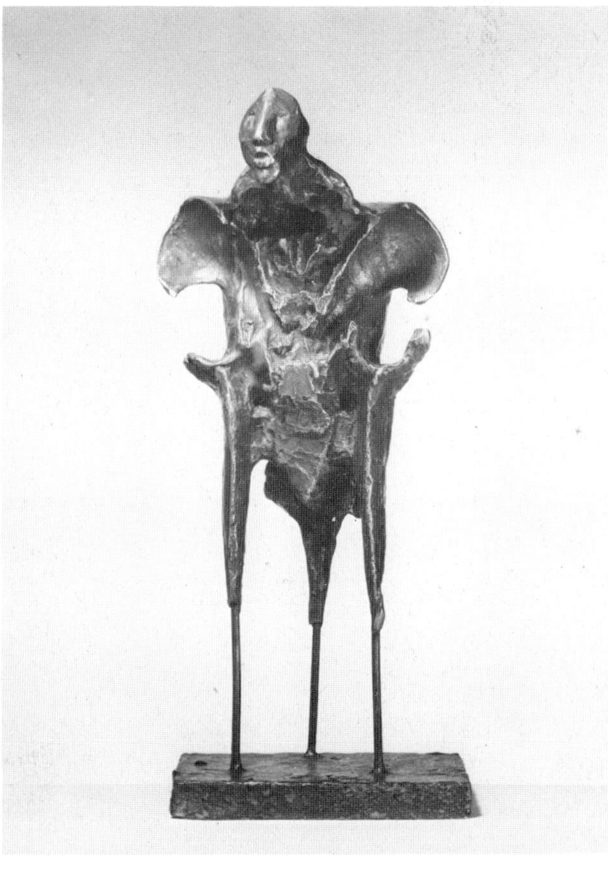

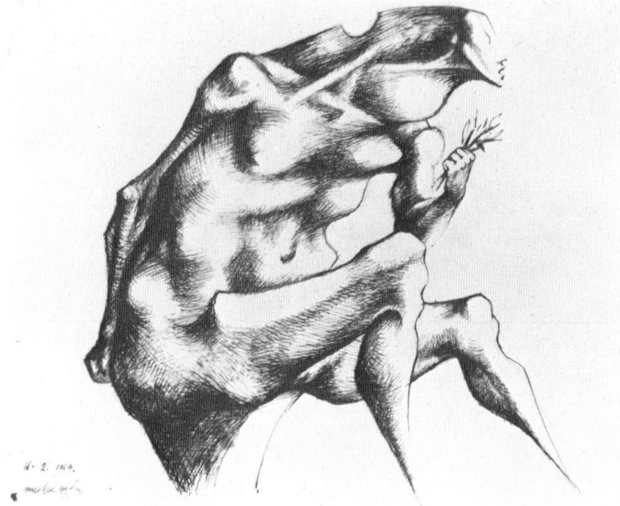

182 *The Oracle as Pythia, 1964*

181 *Mantic Figure II, 1962*
183 *Oracle — Responding Head I, 1963*

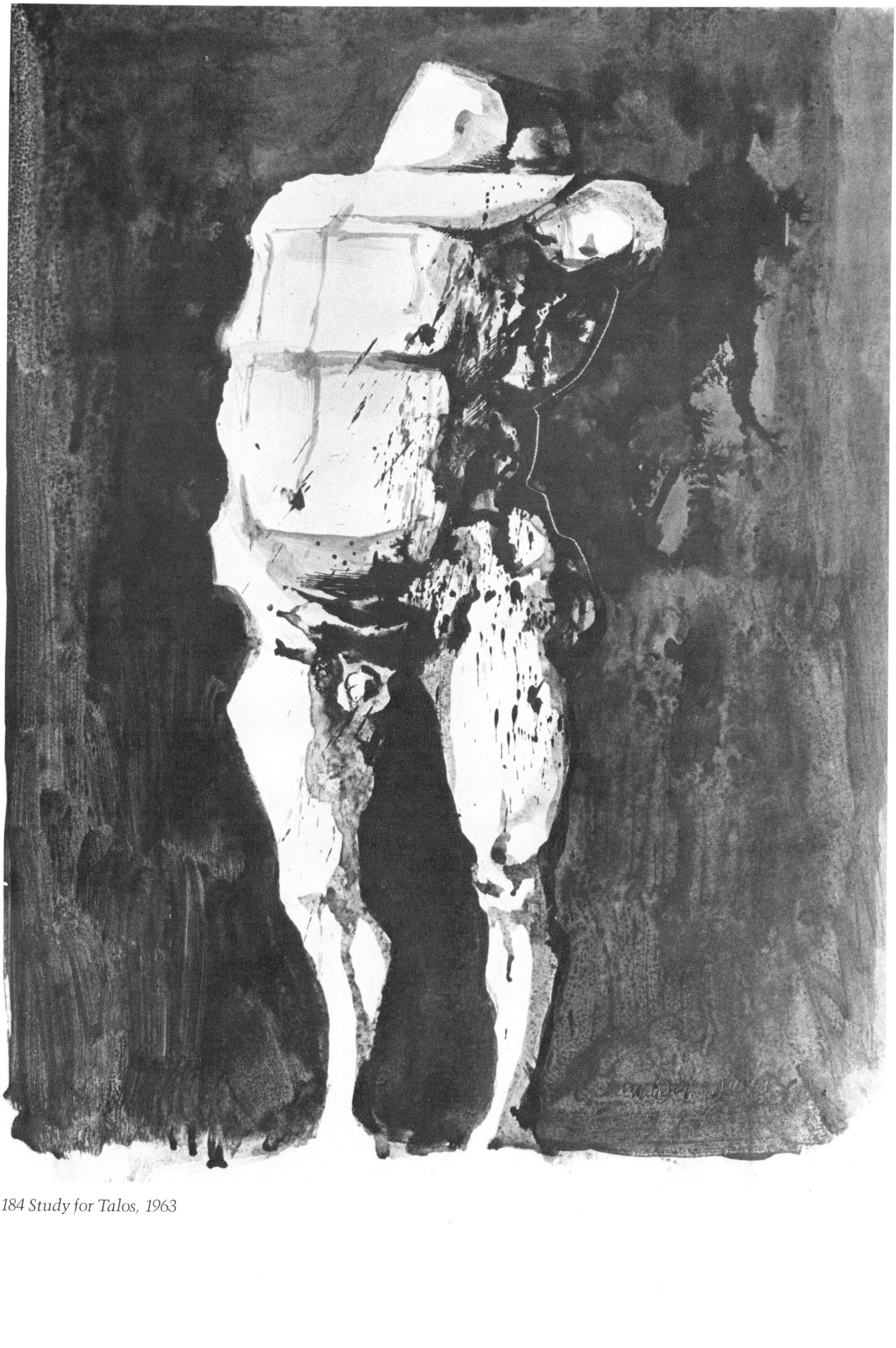

184 Study for Talos, 1963

185 *Demeter and Kore*, 1965

186 Seasons of Demeter I: Spring, 1965

187 Seasons of Demeter IV: Winter, 1965

188 Seasons of Demeter II: Summer, 1965

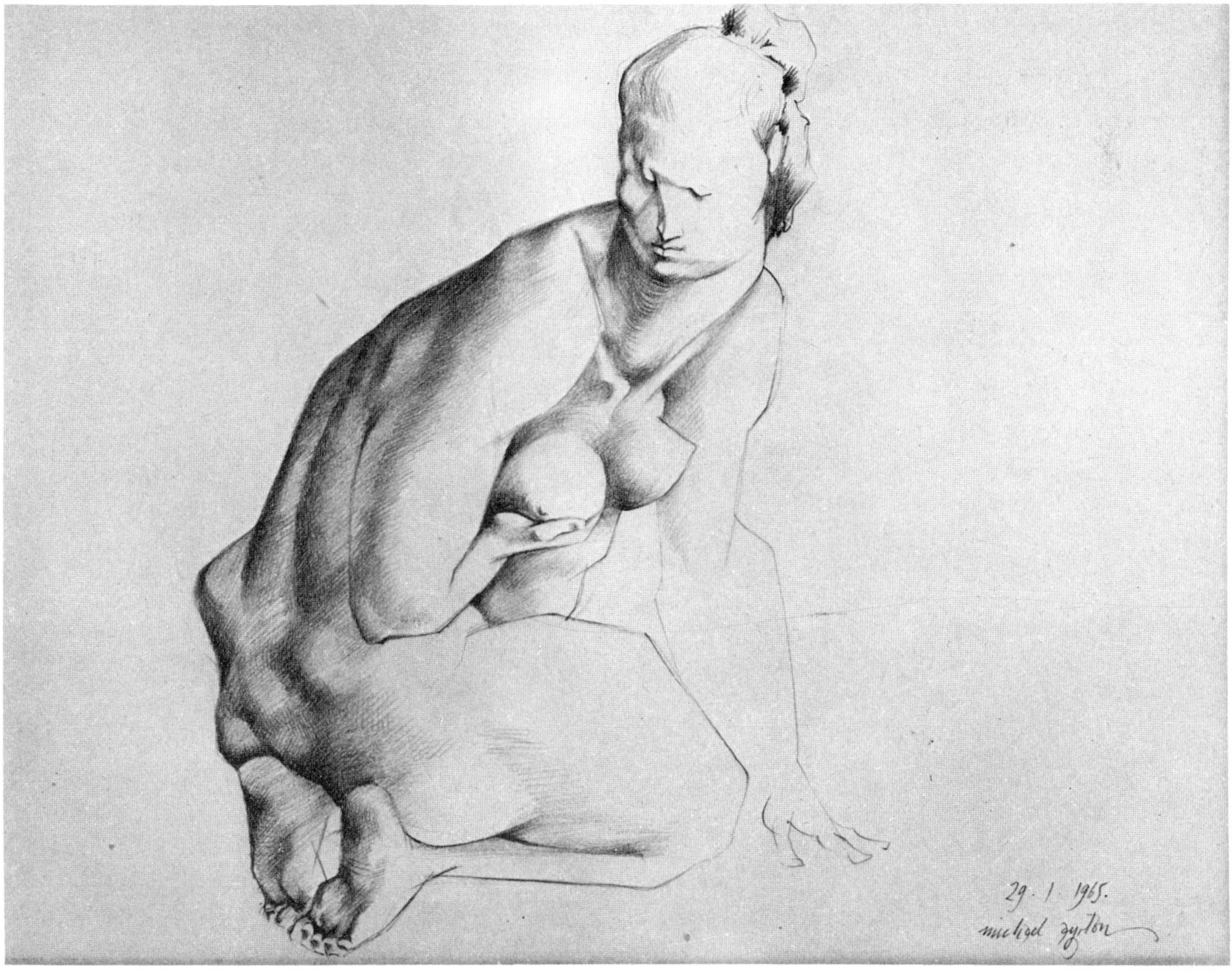

189 *Demeter Pregnant, 1966*

190 *Winter Demeter, 1967*

191 *Seasons of Demeter III: Autumn, 1965*

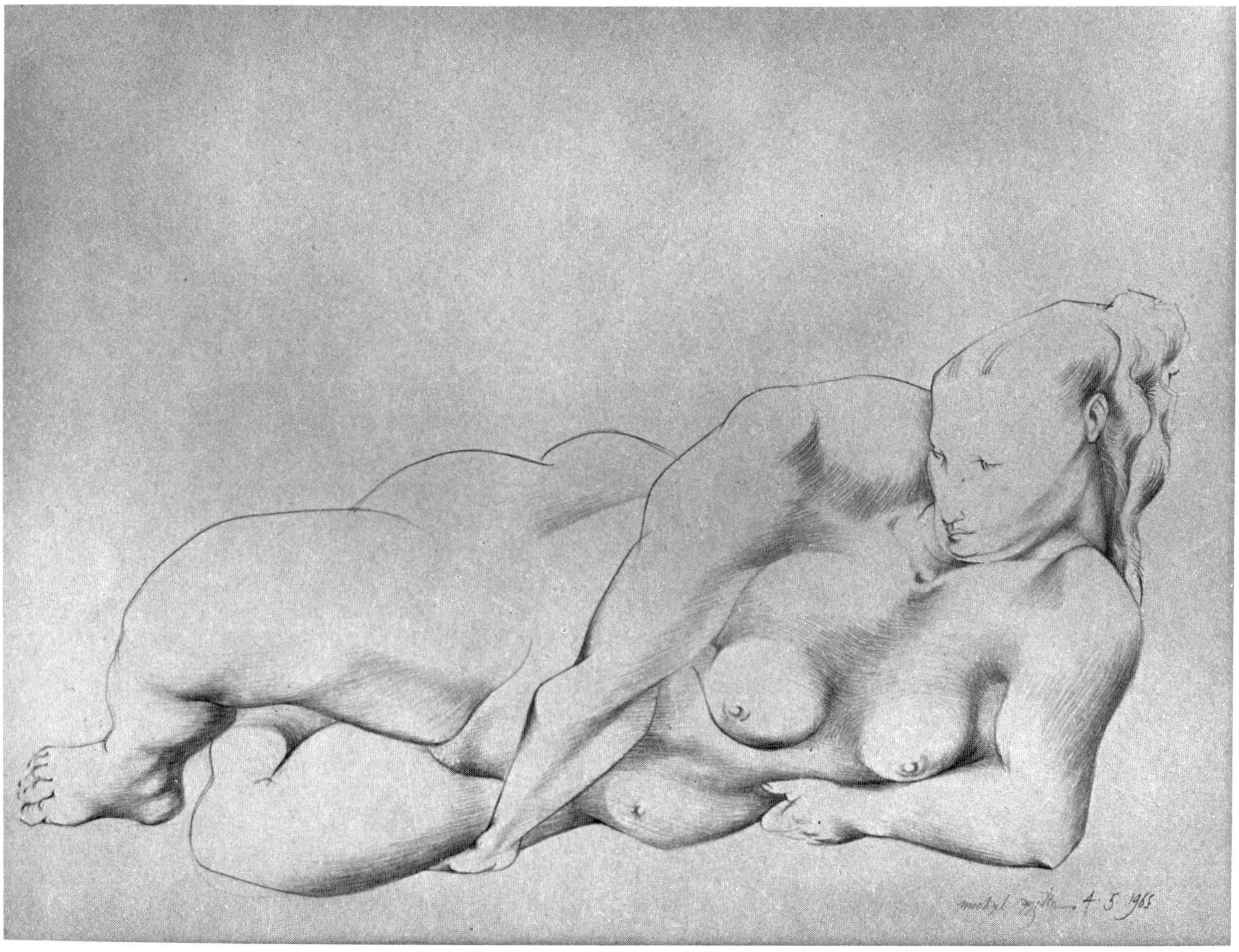

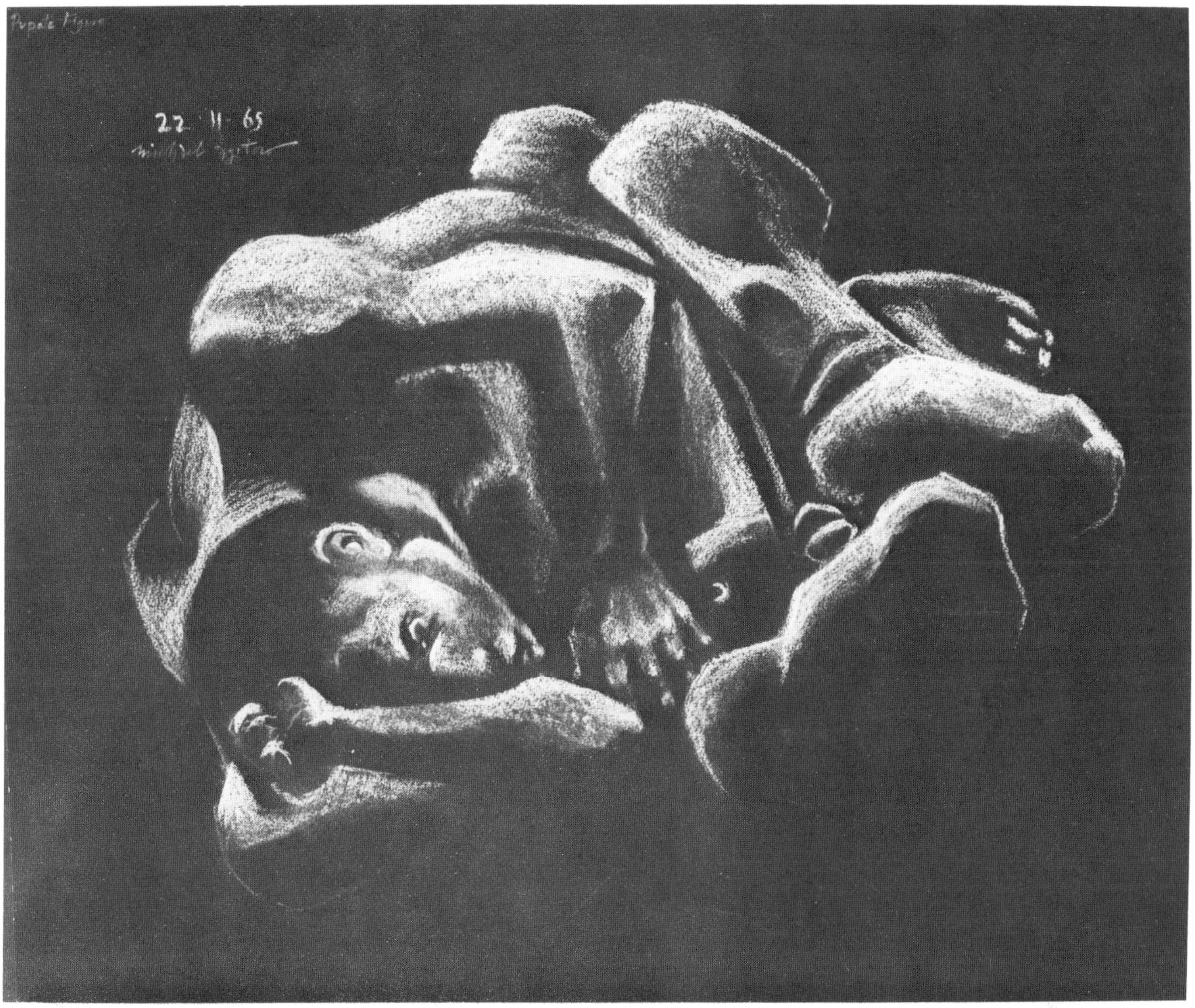

192 Pupate Figure, 1965

A further development of the mid-1960s was concerned with sleeping figures in a more or less foetal position, such as in the powerful white chalk drawing on black paper *Pupate Figure* (1965). This concern with figures in the womb relates back to the maze compositions, but on the other hand it leads on ultimately to the emerging figures and the reflectors. The imagery is specific in the bronze figure *Return* (1969), despite Michelangelesque echoes, while subsequently in the drawing *Compact Anatomies* (1970) the distortions are more extreme. Unfortunately, the latter was not developed into independent bronzes, but the cross-fertilisation with the Minotaur theme led to the *Minotaur Asleep, Miniature* (1962). Much more important is the End Maze series of bronzes which stem from the projected extension to the entrance/exit of the Arkville Maze. In its design, the Arkville Maze follows the Classical type with decision points and dead ends, and Michael Ayrton had already laid out a unicursal floor maze of the Christian type, based on that on the floor in Chartres Cathedral, adjacent to it. The large scale model was never produced but the imagery of *End Maze II* and *End Maze III,* both of 1970, is explicit with the sectioned nautilus shell superimposed on the 'labrys' or double-headed axe, symbolising the original maze or Knossos. The nautilus, with its complex multi-chambered internal structure and central cord, was often used by Michael Ayrton as a metaphor for the maze and the cord of Ariadne and, although in *End Maze II* the emerging figure stands in front of the labrys-nautilus structure, in *End Maze III* the two elements in the composition are amalgamated and, partly enclosed within the nautilus shell, the figure frantically struggles to break out from it. The expanded concept is revealed well by the preparatory drawing, *Study for End Maze III,* while in *Carapace* (1972) the figure enters the nautilus/Maze and is progressively transformed by it. Thematically, though not formally, linked to the End Maze series is also *Point of Departure* (1970) which, like *Personal Janus* of 1970, states clearly Michael Ayrton's personal involvement with the myth. On one side of the wall, identified by the labrys as the Maze at Knossos, Icarus is ascendant over the Minotaur while on the other the sculptor searches for a way through the Maze.

193 End Maze II, 1970

194 Return, 1969

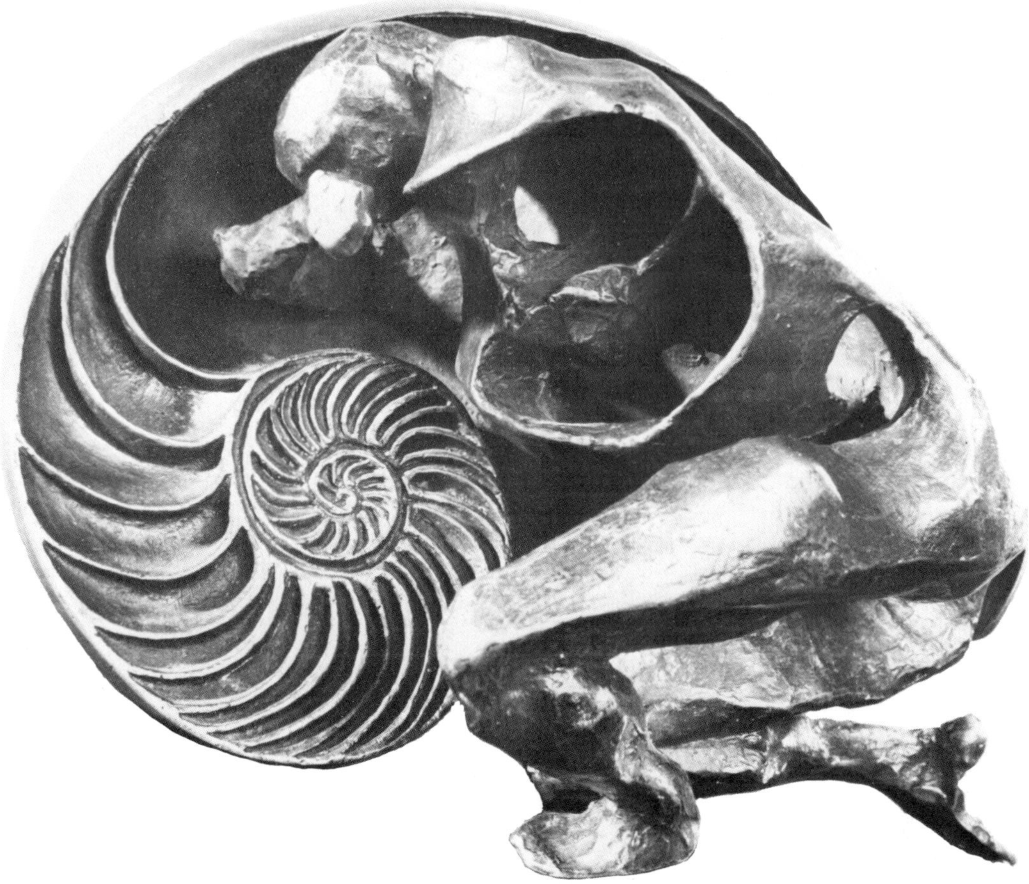

195 Carapace, 1972

196 Study for End Maze Drawing III, 1970

197 Berlioz in Rome, 1831, 1947

For Michael Ayrton the music of Hector Berlioz was a lifelong passion, in part going back to his stay in France in 1938-39. In 1956, however, he was deeply moved by two broadcast performances of *The Trojans* conducted by Sir Thomas Beecham and the impact of the music was heightened by his travelling out to Italy within the week. Fascinated as much by the man as by his music, Michael Ayrton created for the B.B.C. early in 1969, *Berlioz: A Singular Obsession* as part of the Berlioz centenary celebrations. For this he wrote the script, narrated it, selected the music and produced not only the large number of paintings, drawings and sculptures of Berlioz required but also, for the first time for eighteen years, the costume designs. The critical response to this outpouring on television was understandably mixed, but in the production he set out to create a pictorial biography of the great composer through his memoirs and letters and through Michael Ayrton's own intense personal response to the music. The almost mystical experience of the visit to Cumae in 1956 was due in the first place to Michael Ayrton's love of Berlioz, and, together with his longstanding love of music and intimate contact with Constant Lambert, Sir William Walton, Humphrey Searle and others, the extravaganza of *Berlioz: A Singular Obsession* can be seen to be not an isolated phenomenon but closely integrated with his other interests. By comparison the group of bronzes loosely titled *Couples* presents the opposite extreme. Totally uncomplicated, and executed at irregular intervals when new ideas suggested themselves, these bronzes, such as *Curled Sleeping Figures* (1967), *Couple II* (1969) and *Afternoon* (1973), develop ultimately from his early *Bathers* and the curled figures included in *Figures on a Jetty* (1954-55). The emphasis is on relaxation as opposed to stress, on life, love and sex.

The end of the 1960s was a period of intense activity for Michael Ayrton with a large number of exhibitions and the detailed research required for *Giovanni Pisano, Sculptor*, published 1970, but during these years the lateral developments from the Maze theme became increasingly important. In 1972 Michael Ayrton summed up the evolution of the major group of these sculptures, which can be conveniently assembled under the title of 'Emerging and Penetrating Figures', with the statement:

> 'When I had completed the book called 'The Maze Maker' I thought foolishly that I was done with the image of the maze and the *dramatis personae*, who played their parts in an ancient legend. I was wrong. I found that the final words which I had put into the mouth of Daedalus, as his 'autobiographer', were more imprisoning

198 Berlioz in 1852, 1968

199 *Sleeping Figures*, 1967

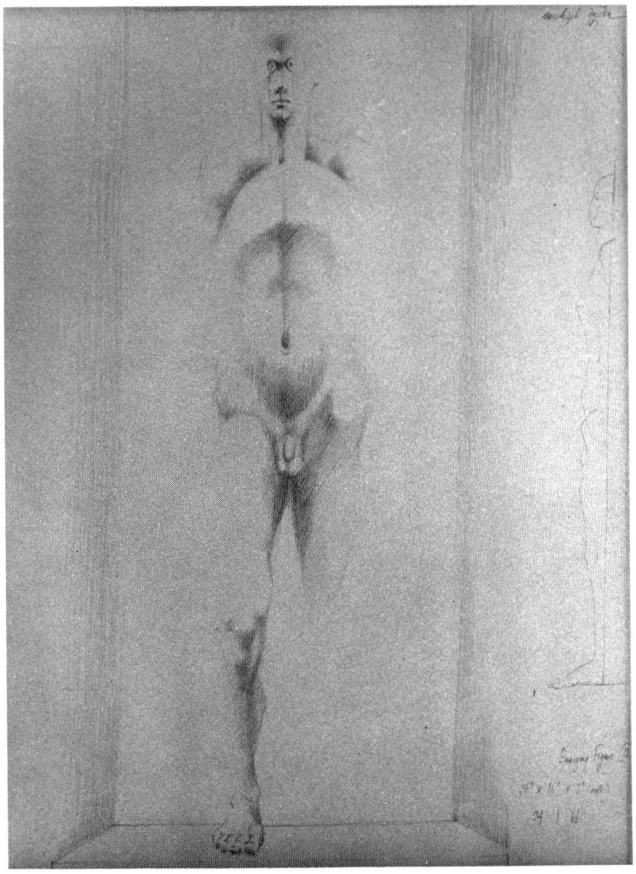

200 *Emerging Figure II*, 1966

201 *Afternoon*, 1973

that the mere termination of a mythical narrative, essentially written to link images I had been making in other media. The sculptures called *Penetrators* and *Extricators* I discovered to be essentially optimistic in that the notion of escape is illusory. The metaphor of the labyrinth is too embracing; to escape from it is to return, wishing to return. My imagery has remained mazed because it cannot be otherwise, being both microcosm and macrocosm and independent of scale. Daedalus called the maze his toy, as well as his trial and his trap, and in this aspect I find that it is sometimes strung with the skein of red thread which Ariadne gave to Theseus, so that my hands are perhaps trapped in that red cord as if I were an inexpert performer on a stringed instrument which produces a music to me no less haunting for being silent.'

The series of bronzes opens with the triptych *Emerging Figures I, II and III* (1966) in which the figure slowly emerges from the neutral ground of the bronze relief only to be increasingly enmeshed. There are clear parallels with the complex painting *Contained Maze* executed in April 1966. In it, the concept of the maze within the man is explored and through the collage elements and stretched threads, that maze is linked visibly to the outside world. The main line of development, however, leads from *Emerging Figures I, II and III* to *Triptych II, Emerging Figure* (1967) and *Through the Blade I and II* (1967 and 1969 respectively). The latter bronzes also develop from the abortive *End Maze* project for Arkville and basically exploit the same metaphor as both *End Maze* and the *Emerging Figures*, except that a shoulder blade is employed as the foil. These *Emerging Figures* preceded the Penetrating Figures, which Michael Ayrton evolved in 1969-70, and reach their climax in *Troy Maze* (1972). This large bronze takes its title from the inscription TRUIA in reverse (mirror image) on the side opposite to the figure (in punning reference to the first four letters of the artist's name), and is preceeded by a particularly fine series of pen and ink studies.

In the Arkville Maze, completed in 1969, the central chamber containing the largest version of *Daedalus/Icarus Matrix II* is, as noted before, lined with reflective bronze sheets as mirrors. The complex multiple reflections introduced the rich web of spatial ambiguities which had first attracted Michael Ayrton in 1966 with the construction of *Mirror Maze.* In certain respects this highly important seminal composition develops out of the Emerging Figures and the painting, *Contained Maze,* of the same year, and is closely related to the earlier maze and stringed figures in its exploitation of the red thread/umbilical cord motif, but the enclosure of the composition within five reflective surfaces formed of bronze sheets makes it infinite in extent. In another statement made in 1972, Michael Ayrton gave an account of how *Mirror Maze* and the Arkville Maze provided the impetus which led to the Reflectors:

'The labyrinth is at once ambiguous in its form and in its nature. To Daedalus its walls were dense and impenetrable, translucent and

202 Contained Maze, 1966

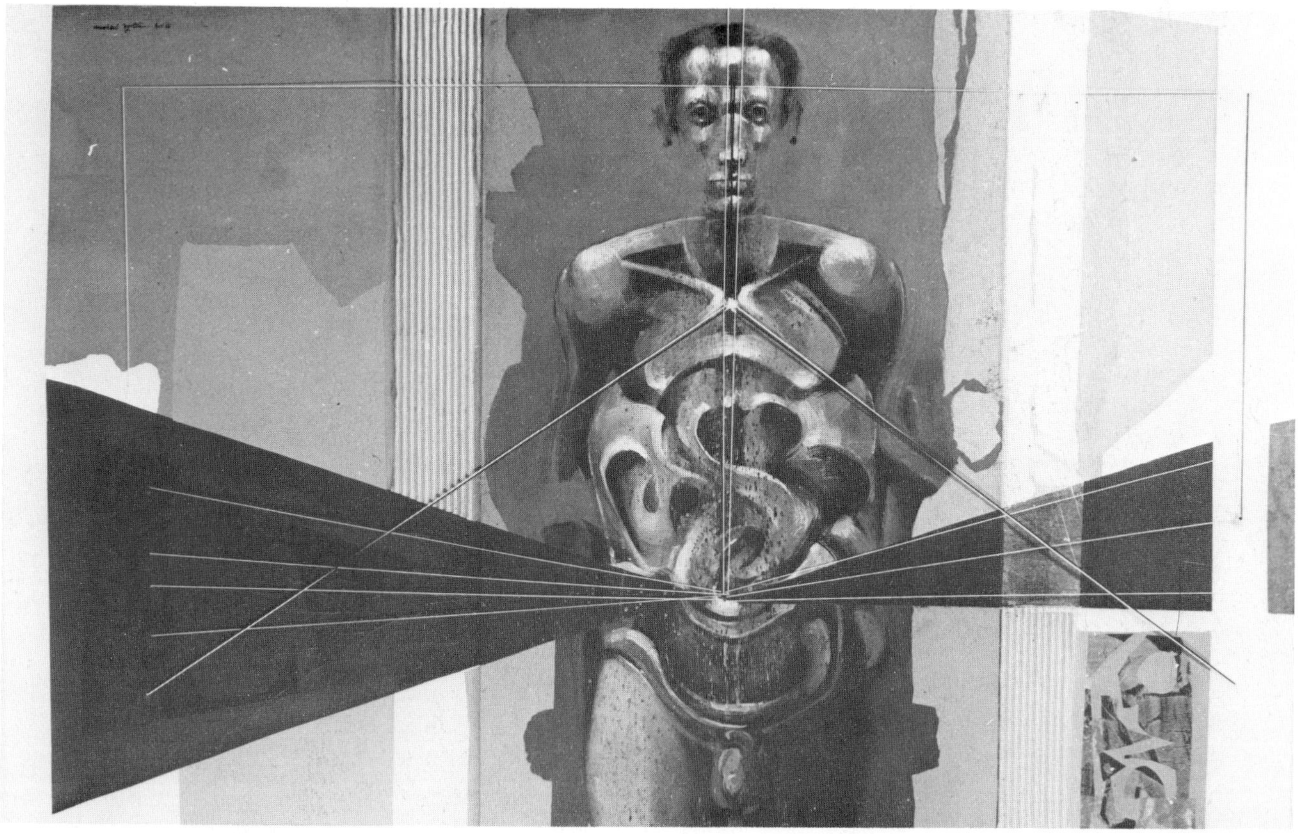

203 Triptych II Emerging Figure, 1967

204 Through the Blade II, 1969

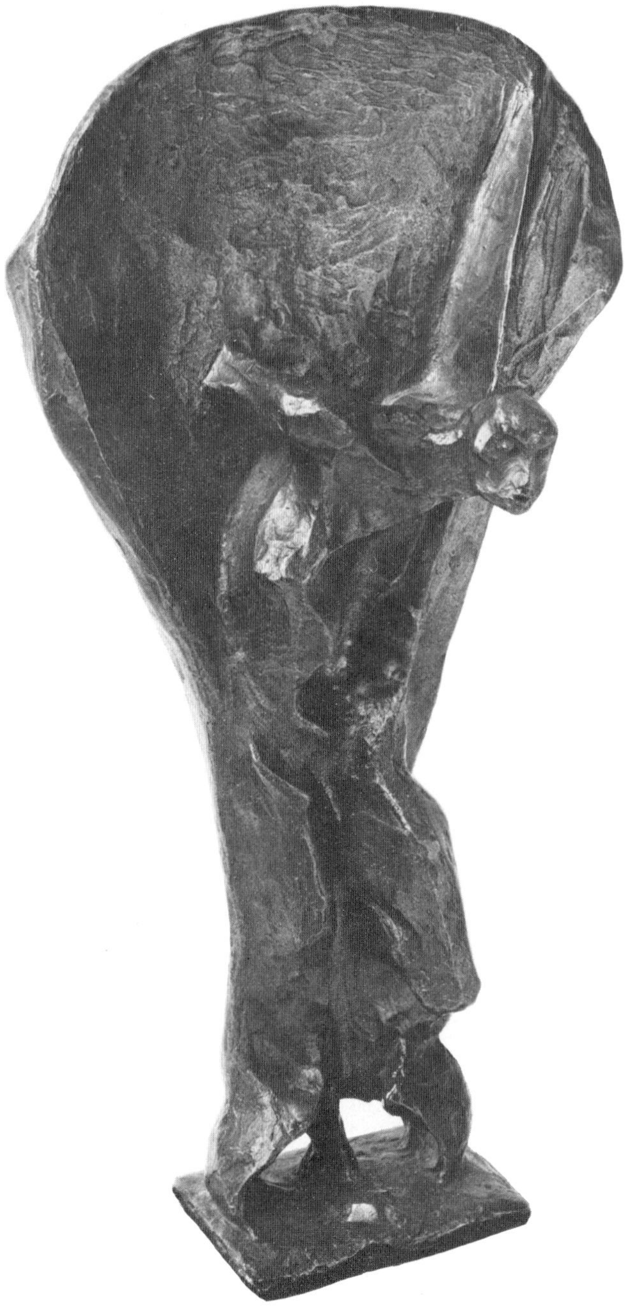

illusory. It may also be as implacable as stone or as impalpable as smoke. As a metaphor it developed for me as *Mirror Maze* in 1966 and as the facetted walls of the innermost coil of the Arkville Maze in 1968. These walls throw back the reflection of the bronze *Daedalus* and *Icarus Matrix* distorted. In 1969 a small piece of neutral perspex, a substance at once dense and translucent yet with the power to reflect, extended my means to give physical expression to metaphorical form as to the idea of what lies within or beyond the mirror. The idea itself is not new, except perhaps in sculpture. Lewis Carroll took Alice through the looking-glass and Jean Cocteau conceived, on film, the means to take Orpheus through the mirror and into the nether world. In ancient legend, it was sometimes believed that a man's image, reflected in clear water, was possessed of a separate spirit from his own, whilc all of us see ourselves in the looking-glass not as we are but in reverse, a maze concept because, although the mirror reflects the image and may even suggest the journey, it gives no clue to its destination. The mystery of the mirror was defined in 10th century China when it was pronounced that at one hundred yards the mirror can see the man, but the man cannot see the mirror and also that the mirror can exist without the image but is incomplete because it is incomplete because it is alone, whereas the image cannot exist without the mirror, but is not itself empty. This riddle can be resolved and perhaps has been, in a sculpture 22 feet high, now in Detroit, and the solution may be imagined in terms of the model for it. On the large scale, each 9 ft. 6 ins. head is half illusory, being completed in reflection, and each large head contains a smaller interior head which is duplicated by the double mirror. Thus the image cannot exist without the mirror which is not itself empty'.

205 *Penetrator II, 1969*
206 *Penetrator, 1970*

207 Troy Maze, 1972
208 Troy I, 1972

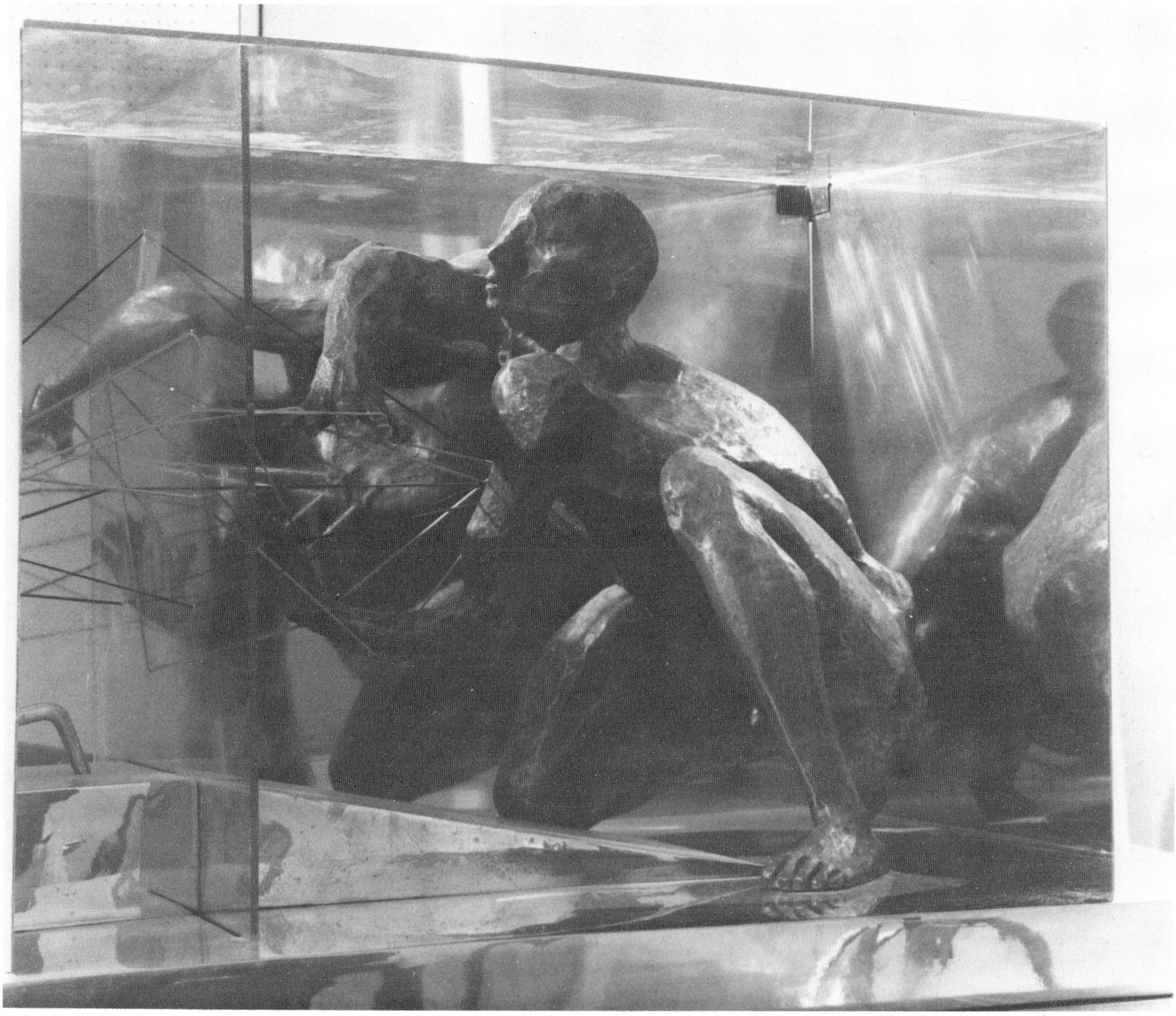

209 Mirror Maze, 1966

Comparisons drawn between Michael Ayrton's reflectors and stage effects achieved by the use of large sheets of polished plate glass set obliquely to the audience are of limited significance because the control over the lighting of the background must be entirely separate from that of the object whose reflected image is seen by the audience if a convincing illusion is to be achieved. The new development central to Michael Ayrton's reflector sculptures was the replacement of the transparent plate glass with neutral density perspex. Under uniform lighting conditions neutral density perspex possesses a special optical property in that the image transmitted through the smoky sheet is of the same intensity as that reflected off the surface. Thus in the simplest of reflectors, such as *Mirrored Figure* (1972), the eye receives two images—that of the polished bronze figure itself and that reflected from the surface of the sheet which suggests the existance of a second figure on the other side of the sheet (the virtual image). As *Mirrored Figure* is turned around its vertical axis, by means of the turntable built into its base, the real image and the virtual image maintain a simple relationship to

210 Mirrored Figure I, 1970

one another, but in *Dioskouroi* (1970) and *Mirror Twins* (1973) the full properties of neutral density perspex are brought into play. A second polished cast of the same figure in bronze is placed in an identical position on the opposite side of the sheet of perspex so that now the eye receives three images: that of the polished bronze figure on the side closest to the observer, the virtual image of that figure obtained by reflection and the transmitted image of the second bronze figure seen through the sheet. The use of neutral density perspex ensures that the virtual image and the transmitted image are of the same intensity and when the reflector is turned on its vertical axis a complex changing interplay between three images is obtained. What could be merely an interesting scientific toy is, in the hands of Michael Ayrton, the means by which he challenged the generally accepted limits of reality. The dynamic relationship between the virtual and transmitted images as the more complex reflectors are turned is an extension of his earlier sculptural ideas in that they both penetrate and emerge. Heads and figures, and other elements, penetrate each other and then emerge, as in *Reflex II* (1969) and *Kouros* (1969), and the potent imagery of the maze is reintroduced through its visual metaphor—the nautilus shell—in *The Discovery of Nautilus* (1969). In *Entrance* (1970) Michael Ayrton took as his point of departure *Blade Figure II* while in *Brain Maze* (1972) and its related drawings the interactions are particularly subtle and well-defined.

211 Dioskouri, 1970

212 The Discovery of Nautilus, 1969

213 *Mirror Twins, 1973*

All the reflectors exploiting the optical properties of neutral density perspex must be turned around their vertical axes for the spatial effects to be experienced, but in terms of size there are mechanical limitations on this design and, instead of the sculpture turning, the huge Kresge Head must be explored by walking round it. This twenty-two foot high reflector was commissioned for the outside of the Headquarters of the S.S Kresge Company in Troy, Michigan, and consists of two huge bronze half-head forms divided by a bronze glass screen and set on a high facetted bronze plinth. Set up in October 1972 it remains the largest single sculptural form created by Michael Ayrton and, by means of the smaller masks enclosed within, it was his intention to convey the principle that 'it is in the conjunction of heads that human beings relate and combine to formulate concepts'. In the smaller reflectors, the idea of heads within heads had been explored as early as *Reflex II* (1969) and in the excellent ink and wash drawing *Translucent Heads II* (1970), now in the Detroit Institute of Arts, and *Study for Contained Heads, I* (1970). One line of development looks back to the Maze Heads of 1964-65 and, instead of seeing the inside of the head as just a metaphor for the complexities of the Maze, Michael Ayrton experimented with skeletal interpenetrating heads, entitled *Constructed Heads* (1972),

214 *Study for The Discovery of Nautilus II, 1970*

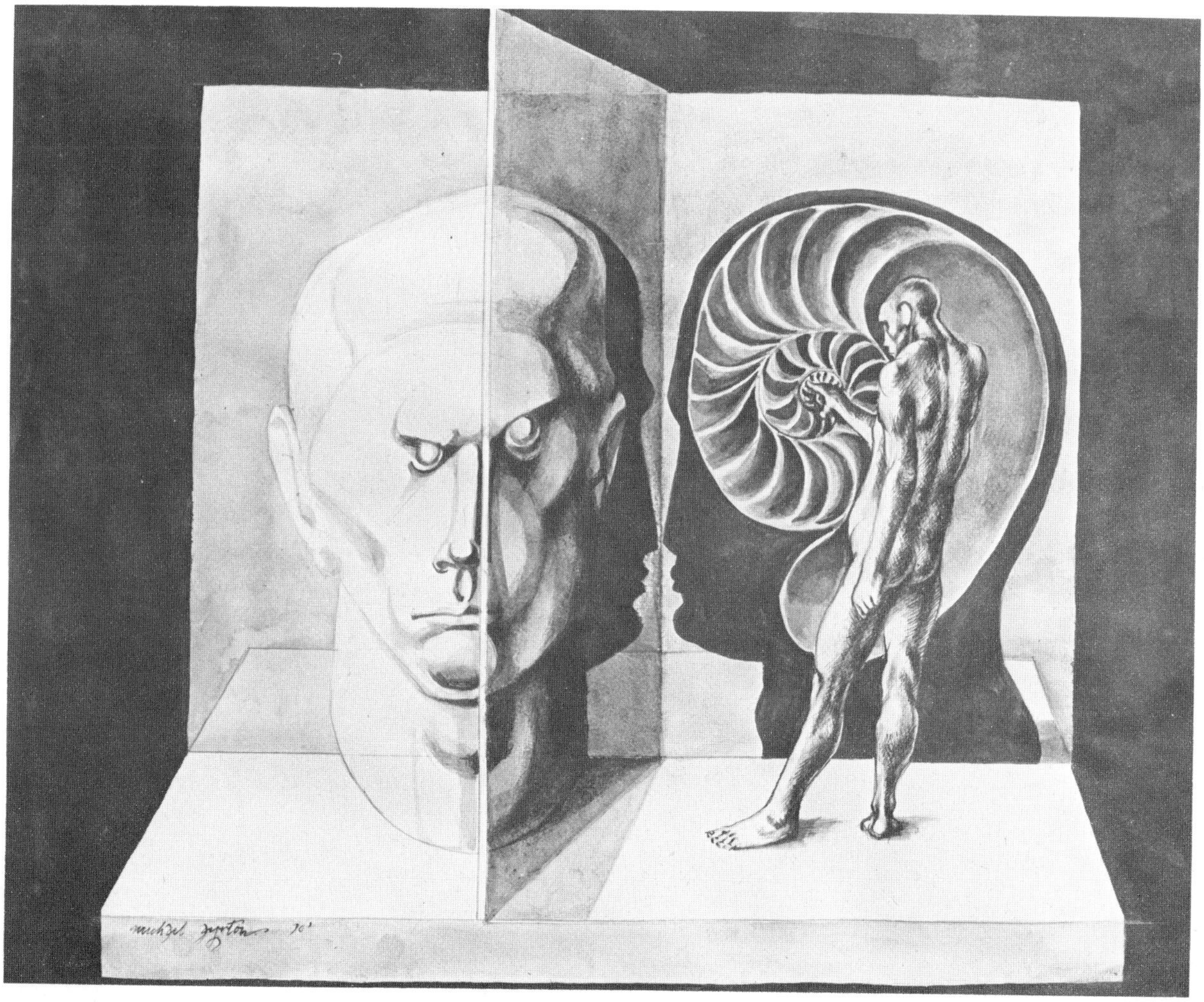

which bring together the *Maze Heads* and the *Contained Heads.* On a small scale, these ideas are also the basis of the bronzes *Captive* (1962) and *Contained Heads* (1970), and by lateral extension lead to *Emerging Head* (1970) and *Tricephalic* (1972). Compound heads with three or more faces, as distinct from the Janus type, are well-known in Renaissance art, and the Titian so-called *Three Ages of Man* in the National Gallery, London, is a familiar example, but the combination of *Tricephalic* within a cut-away head to create *Contained Heads* is also indebted to Henry Moore's *Helmet Heads* with interchangeable inner elements. Jacob Nyenhuis has suggested that the inclusion of the tricephalic deity implies the many aspects of every man's experiences, and, as he reminds us, Daedalus states in *The Maze Maker:* 'the topology of my labyrinth remains ambiguous. Its materials are at once dense, impenetrable, translucent and illusory. Such a total maze each man makes round himself and each is different from each other, for each contains the length, breadth, height and depth of his own life.'

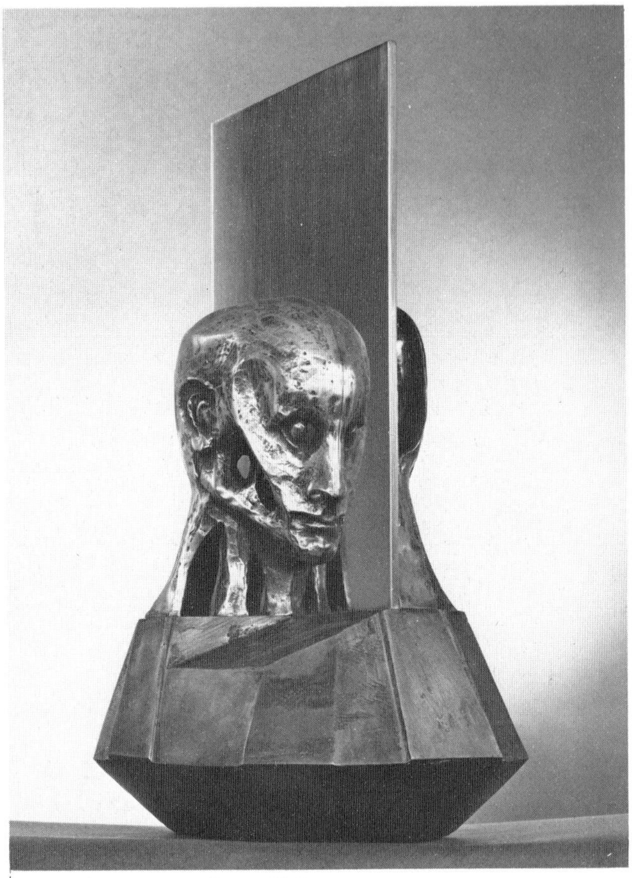

215 Reflective Heads II, 1971

216 Brain Maze (Drawing III), 1972

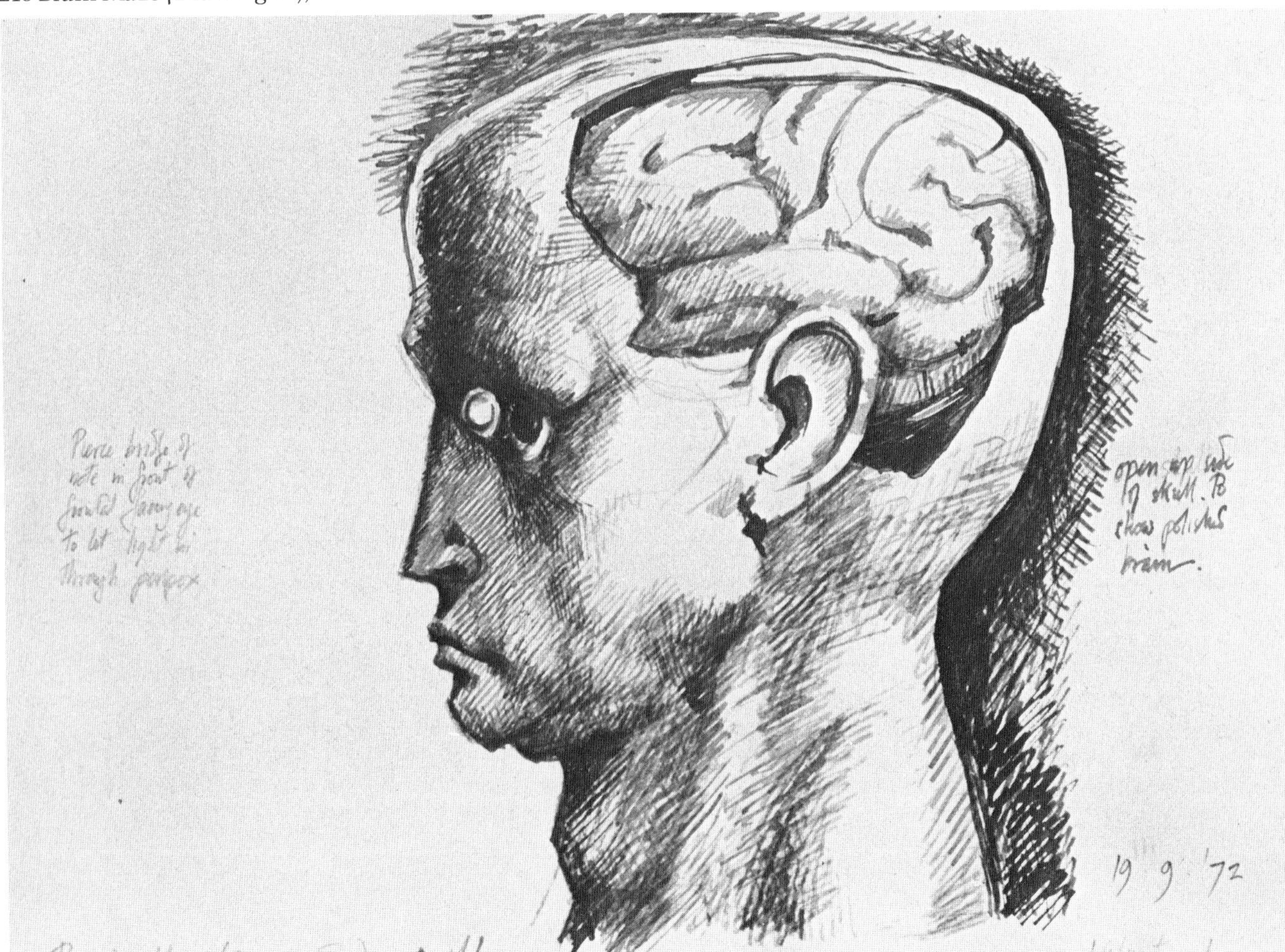

217 Reflex II, 1969

218 Brain Maze IV, 1973

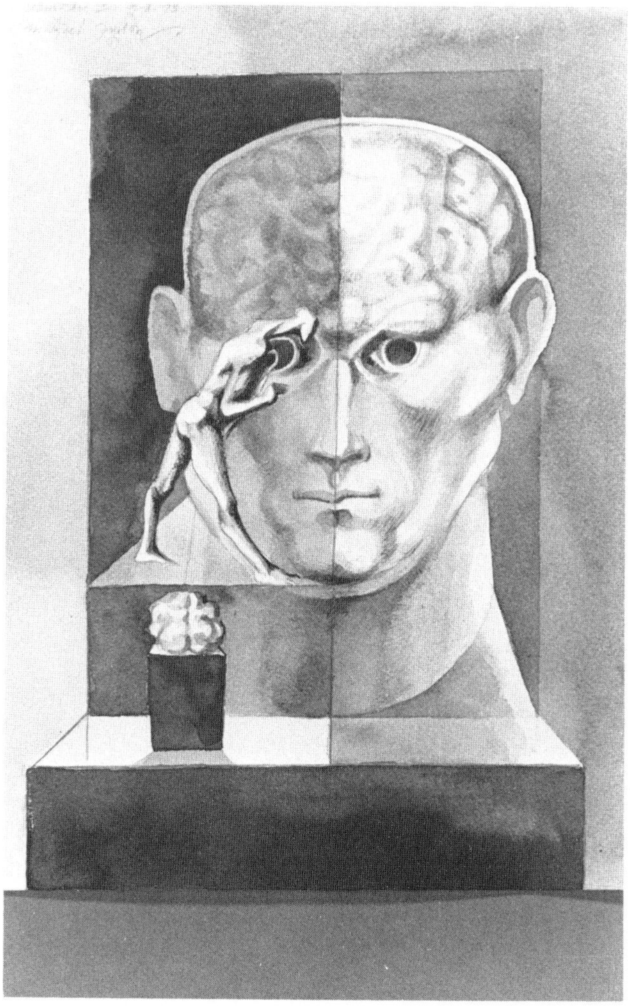

219 Brain Maze, 1972

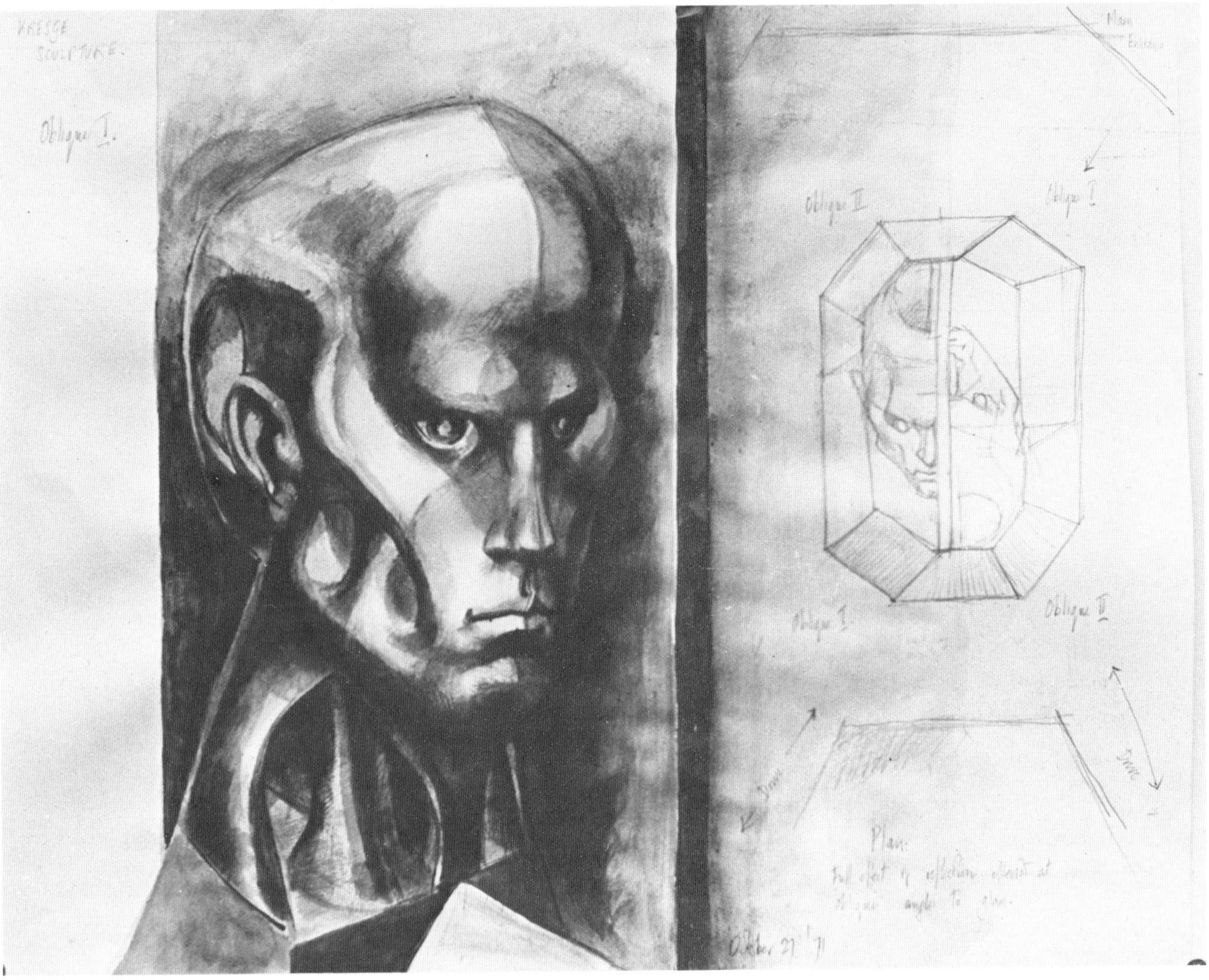

220 *Study for Reflective Head, 1971*
221 *Study for Reflective Head, 1971* (right) 222 *Reflective Heads, 1971/72*

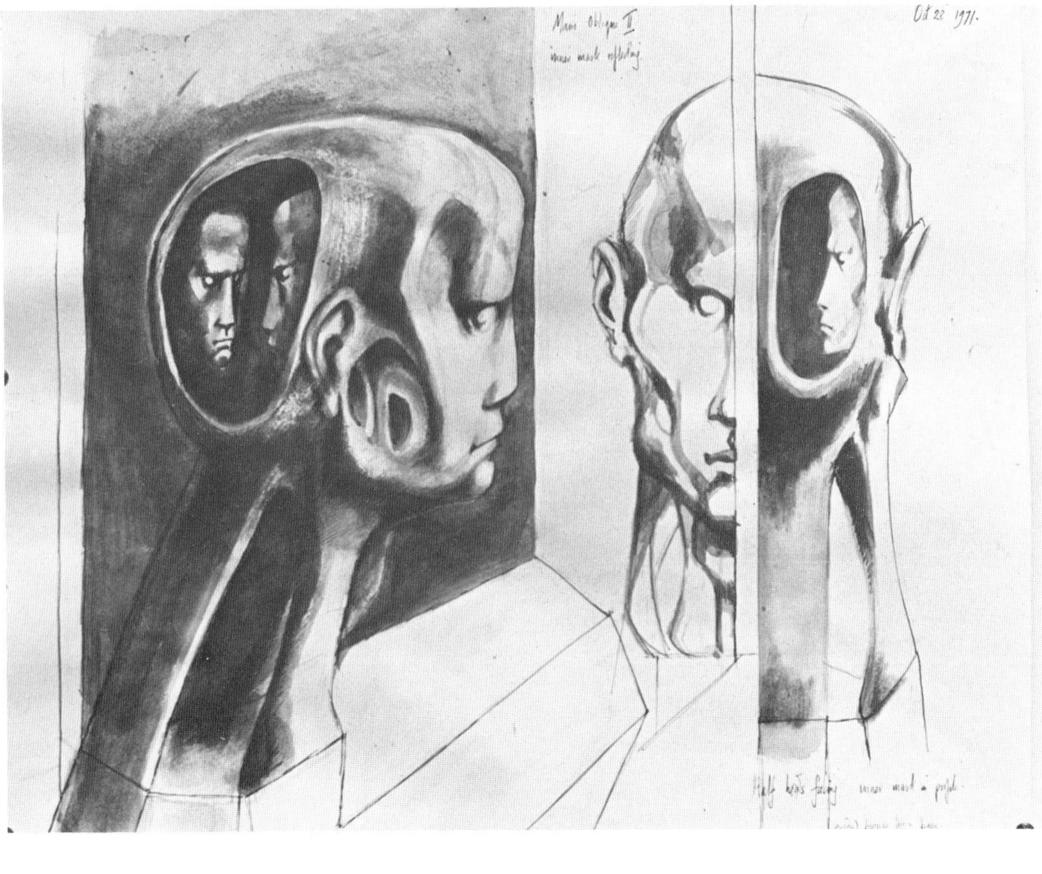

223 *Translucent Heads II, 1970*
224 *Study for Contained Heads I, 1970*

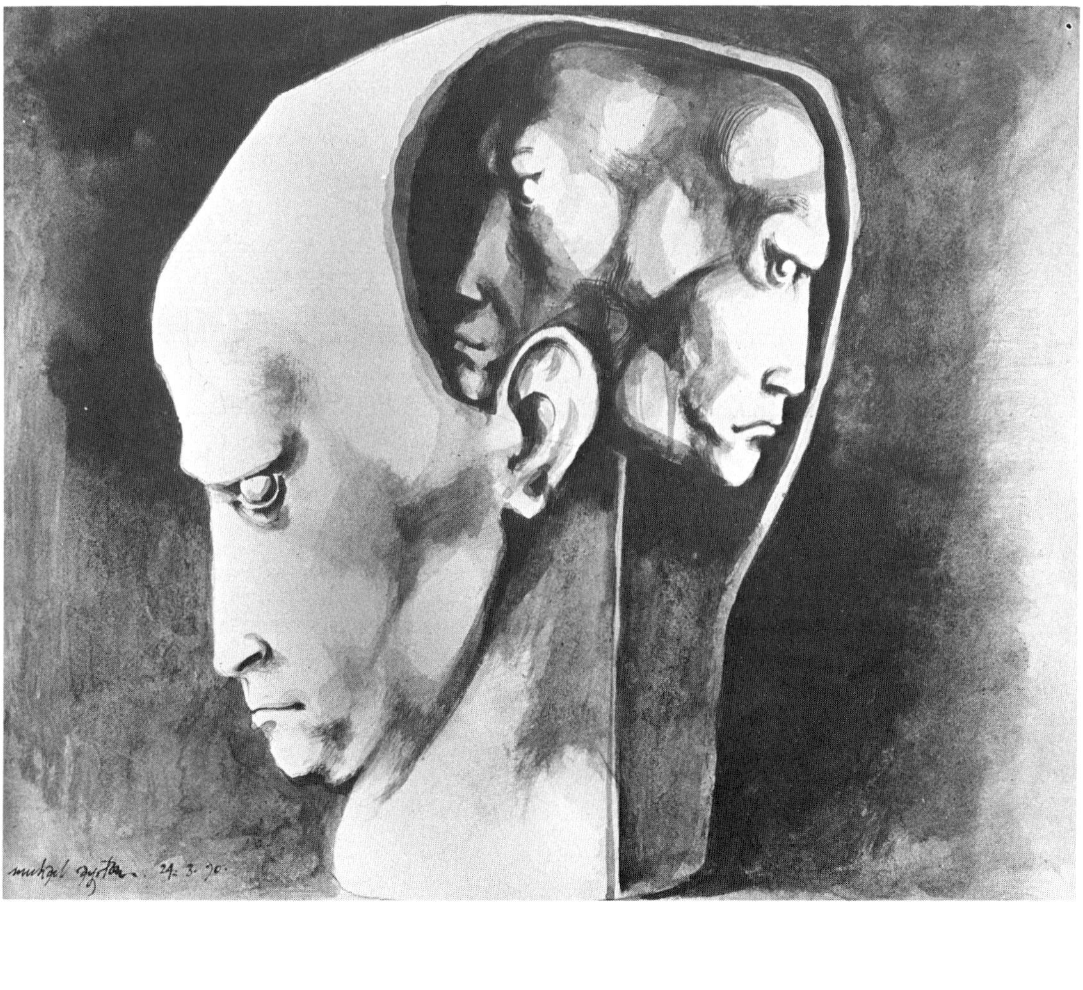

225 Maze Head, 1965
226 Maze Head, 1965

227 Constructed Heads, 1972

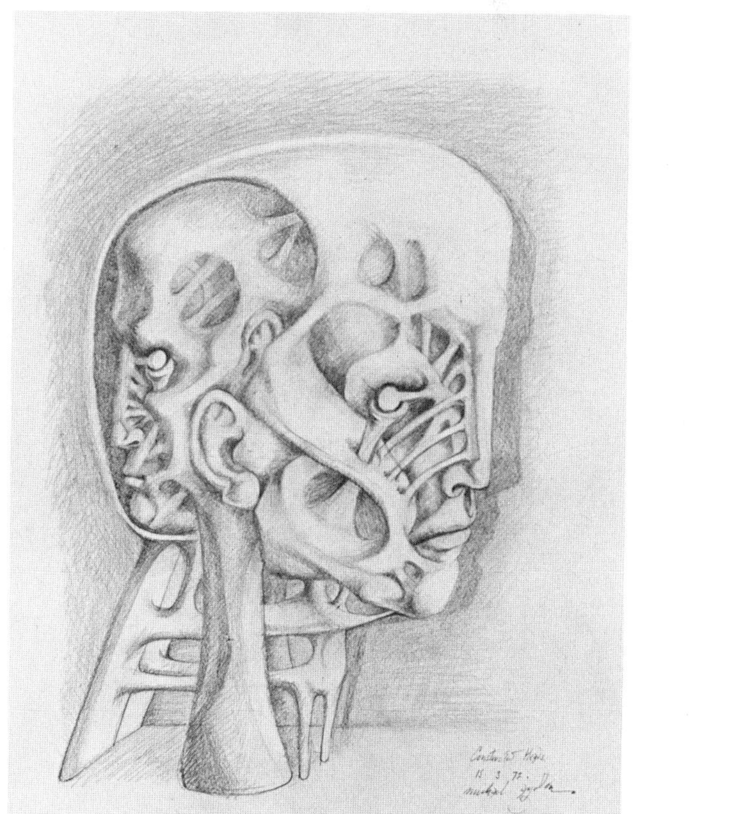

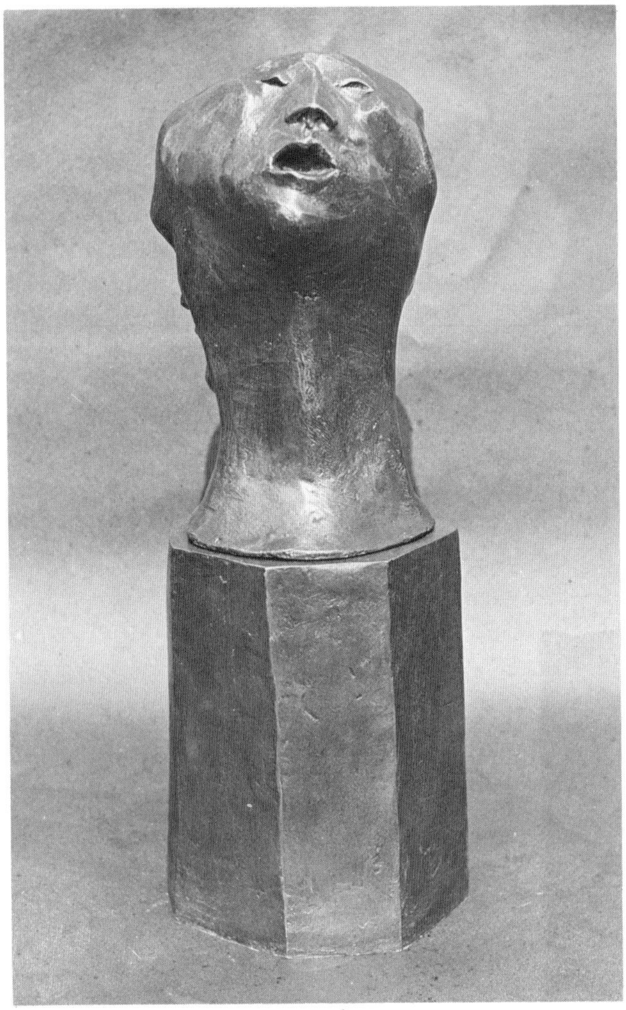

228 *Tricephalic, 1972*
229 *Tricephalic, 1972*

230 Maze Music Drawing II, 1972

Further lateral extensions from the maze theme during the early 1970s include *Red Thread* (1972) in which Michael Ayrton picked up again the idea of bone as a visual metaphor for landscape which had been the foundation for *Orpheus* (1957) and *Orpheus Seeking, I* (1958). In *Orpheus Seeking II* executed in 1971 the fragment of a pelvic girdle is replaced by the entire structure, but the relative size of the figure is larger and the composition is considerably less pictorial in its conception. In *Red Thread,* like *Figure in a Landscape* (1957), the bone elements are used more as metaphors within the landscape than metaphors for the landscape itself, but the addition of the red thread of Ariadne which enmeshes the figure and binds her to the landscape is an extension of the maze predicament. *Maze Music* (1972) is again developed from an earlier sculpture, *Maze Player* executed in 1965, which in turn evolved from the Maze Maker bronzes also of 1965. In the sequence of bronzes, the maze is transformed into a musical instrument and the final imagery is rather more explicit in the sepia ink drawing *Maze Music, Drawing II* which is dated 30.10.72. Nevertheless, Michael Ayrton further explored the ideas of the enmeshed figure in *Web* (1972), where the maze is an invisible trap, and the *Laocoon Maze Figures* (1972) for whom the snakes come from within themselves. In the latter group the concept is again more clearly evident in the series of pen and ink drawings than in the finished bronzes themselves.

231 Maze Music, 1972

232 Red Thread, 1972

233 Study for Web, 1973

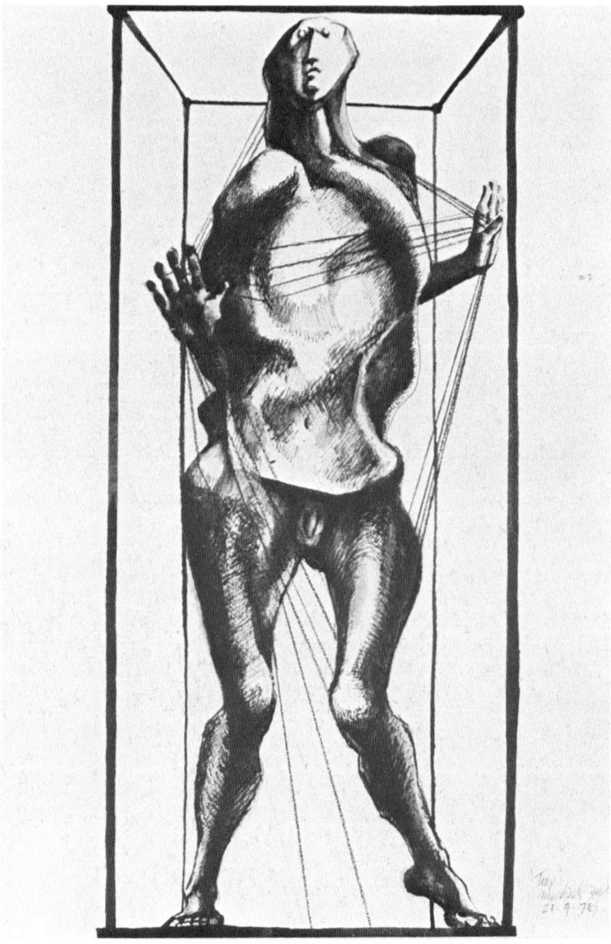

234 Web, 1972

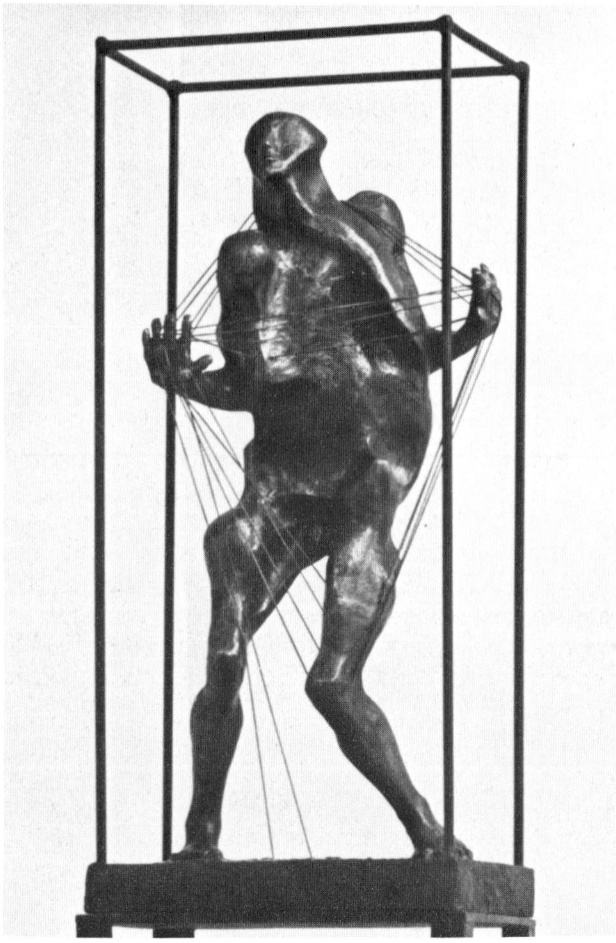

235 *Laocoön Maze Figure III, 1972*
236 *Laocoön Maze Figure II, 1972*

In many respects 1974 was a lost year, although a number of sculptures were finished then and bear that date, and the ravages of undiagnosed diabetes very nearly killed Michael Ayrton. From April he worked slowly on the six different models for his *Chess Set* and the laborious finishing of the thirty-two waxes required for each set as he struggled against the disease and his advanced arthritis. Only in July 1974 did he feel able to inscribe on the drawing *Transformation:* 'To my surprise, in September 1973, a head, in wax, based on a memory of a photograph of the mummy of the Pharaoh Seti I became a recollection of the self portrait of the architect Leon Battista Alberti. This drawing records the transmutation.' Later in the year he modelled the head of *Alberti,* inspired by the famous plaquette in the Kress Collection, Washington, but much stranger and more personal is the sombre wash drawing *Eclipse* dated October 1974. Although the imagery is inspired by Leonardo da Vinci, Goya's acquatints, and the apocalyptic drawings of Dürer, the inscription: 'After this men can believe anything. They can expect anything. Be not astonished any more although you see beasts of the dry land exchange with dolphins, and assume their peace in the watery pastures of the sea, and the beasts who loved the hills find the ocean's waters sweeter than the bulk of the land', is a translation from the Greek of a fragment of the verse of Archilochos who experienced the total eclipse of 648 B.C. Executed in black ink and collage, the haunting insubstantiality of the spectre looming over the horizon echoes the optical properties of the Reflectors as well as Goya's prints. However, in *Study for Heavy Figure, I* (executed also in October 1974), Michael Ayrton had started to explore the problems posed by massive Michelangelesque reclining figures with the view to a new series of sculptures and the counterbalance to these figures was a second group of drawings, including *Study for a Hanging Figure* dated 14.11.74, also intended to be developed into bronzes. Unfortunately, neither series progressed any further.

237 Alberti, 1974

238 Transformation, 1974

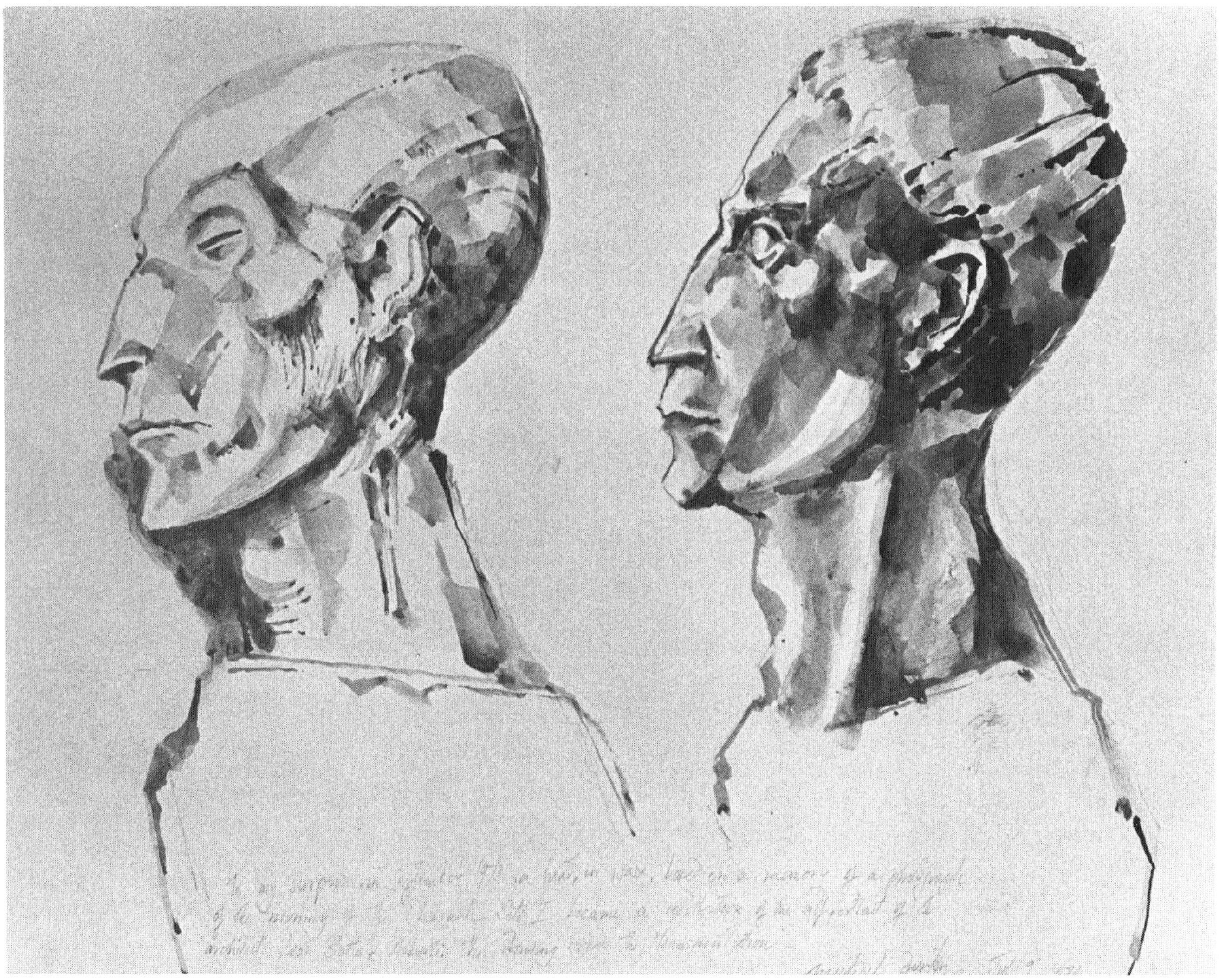

239 Eclipse, 1974

240 Heavy Figure I, 1974

During the Christmas period in 1972-73 the Ayrtons had made their first journey to Africa and to all intents and purposes Michael made no drawings during it. He became, however, passionately interested in the African landscape and the animals which he saw in the game reserves, and on his return he executed two splendid drawings of elephants. The pen and ink drawing on grey paper *Tsavo East* dated March 1973, heightened with coloured pastels, is very unusual in technique for this late date, but it was the first to be done after his return and the treatment of the landscape details is a brilliant evocation of the Tsavo East National Park in Kenya. After his illness in 1974, Michael Ayrton again returned to East Africa as part of his convalescence and in January-February executed a spirited series of animal and bird drawings in ink and wash with the intention of producing a suite of animal etchings. One plate, depicting *Baboons* was finished, but as he slowly regained his strength and old vitality most of his energies were taken up by the research and preparatory work for a series of thirteen programmes on mirrors for television and by the suite of Archilochos Etchings. The television programmes were never completed, but during 1975 Michael Ayrton, assisted by Prue Hopkins, set up an etching studio at Bradfields and sufficiently good progress was achieved as to have the sixteen final plates of the Archilochos suite ready for printing in November. Following the Minotaur Etchings of 1971 and the Femmes/Hombres Etchings of 1971-72, the Archilochos suite finally established Michael Ayrton's position as one of the leading etchers active in England since the War. With the utmost economy, the rapid flowing outlines and lightning movements of the etching needle were softened with varying amounts of drypoint after biting the plates. The starkness of the forms enhances the brutality of the soldiers, evoking superbly the archaic character of the poetry. Although Michael Ayrton himself possessed no knowledge of Classical Greek he took enormous pains to make certain that the interpretation of the surviving fragments of Archilochos's verse was fully in accordance with the most accurate modern translations and the latest scholarly research. On Sunday, 16th November 1975 he entertained Professor Moses Finley and Professor G.S. Kirk at Bradfields to discuss the finer points in the text and afterwards he drove up to London. In his flat there, that evening, Michael Ayrton suffered a massive coronary and died within minutes of reaching the Royal Free Hospital.

241 African Buffaloes, 1975

242 Warthogs, 1975

243 Baboons, 1975

244 Tsavo East, 1973

Impact

Epode

Paros

Spears

245 Suite of Archilochos Etchings, 1975

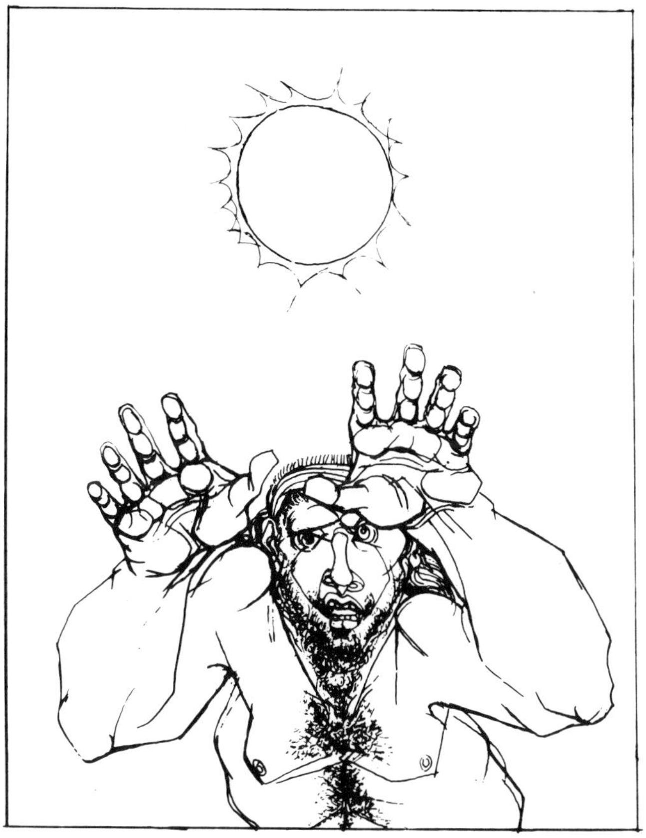

Eclipse

Salmydessos

Lykambes and his Daughters

Archilochos Alone

246 Point of Departure, 1970

List of Illustrations

Self Portrait. Signed and dated December 1946. Brush, pen and black ink, and white chalk. 14¼ in x 10¼ in (Private Collection, London)

1 *The Vase of Flowers.* Signed and dated 29th January 1940. Pen and ink. (Size and whereabouts unknown)

2 *The Bull Ring.* Signed and dated July 1939. Oil on board. 23½ in x 28½ in (Present whereabouts unknown)

3 *Jupiter and Antiope.* Signed and dated February 1940. Pen and ink and gouache. 18 in x 24 in (Present whereabouts unknown; formerly in the collection of James Laver)

4 Michael Ayrton with John Minton examining the *Macbeth* designs, 1941.

5 *The Maskmaker.* Signed and dated February 1940. Pen, black ink and watercolour, heightened with white. 22 in x 16½ in (Calderdale Museums Service).

6 *Strolling Players.* Signed and dated February 1940. Pen and ink and watercolour. (Size and present whereabouts unknown)

7 *War Fortification.* Signed and dated 1940. Indian ink and gouache. 12½ in x 20⅛ in (Victoria and Albert Museum, London)

8 *Fête Champêtre.* Signed and dated July 1940/February 1941. Pen and ink and gouache. (Size and present whereabouts unknown)

9 *The Juggler in Sequins.* Signed and dated 1940. Oil on canvas. 20 in x 24 in. (Present whereabouts unknown)

10 Design for *Macbeth,* Act II. Signed and dated 1941. Pencil, pen and ink and body colour. 10 in x 7 in (Victoria and Albert Museum, London)

11 Design for Malcolm, in *Macbeth,* Act I, Sc. 2. Signed and dated 1941. Pencil, pen and ink and body colour. 10 in x 7 in (Victoria and Albert Museum, London)

12 Set design for *Macbeth,* Act III, Sc. 9, Dunsinane. Dated 1941. Body colour. 14 in x 19½ in (Victoria and Albert Museum, London)

13 *Orpheus.* Signed and dated May 1941. Oil on canvas. 29½ in x 19¾ in (Fay Pomerance, Sheffield)

14 *St. Anthony No. XXI.* Dated December 1942. Scratched into slate. 12¼ in x 11 in (Elisabeth Ayrton, Rockhampton)

15 *Study for St. Anthony.* Signed (1942). Pen and black ink and gouache. 9 in x 8 in (Fay Pomerance, Sheffield)

16 *Study for the Child in the Temptation of St. Anthony.* Signed and dated December 1942. Pen and black ink and watercolour. 14¾ in x 9 in (Fay Pomerance, Sheffield)

17 *St. Anthony's Shadow, 2nd Series IV.* Signed and dated December 1945. Pen and black ink and black wash on waxed paper, heightened with white. 11¾ in x 8½ in (Fay Pomerance, Sheffield)

18 *Temptation of St. Anthony.* Signed and dated 1942/43. Oil on panel. 22 in x 29 in (Elisabeth Ayrton, Rockhampton)

19 *Skull Vision.* Signed and dated 1943. Oil on panel. 25½ in x 32¾ in (Fay Pomerance, Sheffield)

20 *Sketch for Skull Vision.* (Signed and dated 1943. Oil on panel. 5½ in x 7 in (Fay Pomerance, Sheffield)

21 *Joan in the Fields.* Signed and dated 1943. Oil on panel. (Size and present whereabouts unknown; formerly in the collection of Anthony Bushell)

22 *The Earthbound.* Signed and dated 1944. Oil on canvas. 20 in x 30 in (Peter L. Moldon, London)

23 *The Sleeper in Flight.* Signed and dated 1943. Oil on panel. 30 in x 40 in (Present whereabouts unknown, formerly in the collection of Robert Donat)

24 *Convulsive Landscape.* Signed and dated June 1944. Pen and black ink. 13¾ in x 18½ in (Fay Pomerance, Sheffield)

25 *Tree Forms.* Signed and dated 1944. Pen and black ink and black wash with watercolour and coloured chalks. 16¼ in x 23¼ in (Private Collection, England)

26 *Apple Trees by Moonlight.* Signed and dated April 1945. Pen and black ink and wash on waxed paper heightened with white. 20 in x 18 in (Present whereabouts unknown; formerly in the collection of W.E. Evill)

27 *Mrs. Chittenden.* Signed and dated July 1943. Oil on canvas. 20½ in x 24½ in (Private Collection, London)

28 *Entrance to a Wood.* Signed and dated May/June 1945. Oil on board. 30 in x 37 in (Arts Council of Great Britain)

29 *Winter Stream.* Signed and dated March 1945. Oil on board. 22½ in x 32¾ in (Aberdeen Art Gallery)

30 *The White Country.* Signed and dated 1945. Oil on canvas. 40 in x 30 in (Present whereabouts unknown; formerly in the collection of Sir Barnett Stross, M.P.)

31 *The Forge.* Signed and dated November 1945/February 1946. Oil. (Size and present whereabouts unknown)

32 *The Tip, Hanley.* Signed and dated 1946. Oil on panel. 20 in x 24 in (Elisabeth Ayrton, Rockhampton)

33 *Headland, Storm Blowing Up.* Signed and dated 1946. Oil on panel. 20 in x 16 in (Present whereabouts unknown; formerly in the collection of Samuel Courtauld)

34 *Haytor Quarry.* Signed and dated 1947. Oil on canvas. 20 in x 24 in (Lady E.M. Lush, Abergavenny)

35 *Ebb Tide.* Oil on canvas. 10 in x 20 in (Private Collection, London)

36 *The Quay, Isle of Dogs.* Signed and dated 1947. Oil on panel. 10½ in x 12½ in (Calderdale Museums Service)

37 *Thames Foreshore.* Signed and dated 1947. Oil on board. 22½ in x 25¼ in (Students Union, University of Leeds)

38 Design for Head of Mephistopheles, in *The Damnation of Faust.* Dated 1947. Pen and black ink and black wash over pencil. 11 in x 9 in (Private Collection, London)

39 Design for Backdrop 'The Wood' for *The Fairy Queen,* Act I. Signed and dated 1940. Body colour. 15½ in x 21¼ in (Elisabeth Ayrton, Rockhampton)

40 Design for Mephistopheles, in *The Damnation of Faust.* Dated March 1947. Pen and black ink and gouache with fabric swatches. 11 in x 9 in (Private Collection, London)

41 Design for Oberon, in *The Fairy Queen.* Signed and dated 1946. Pen and black ink and body colour. 10½ in x 7⅜ in (Victoria and Albert Museum, London)

42 *The House, Ischia.* 1947. Oil on panel. 14 in x 12 in (Present whereabouts unknown)

43 *Afternoon on Ischia.* Signed and dated 1947. Oil on canvas. 32 in x 53½ in (British Council, London)

44 *Approaching Thunder.* Signed and dated February 1948. Oil on canvas. 34 in x 44 in (Junior Common Room, Exeter College, Oxford)

45 *The Sudden Shower.* (?1947/8). Oil on canvas. 10 in x 14 in (Hollis Baker, Grand Rapids, Michigan)

46 *Conversation in the Wind.* Signed and dated 1948. Two-colour lithography; published by Redfern Gallery, London

47 *Full Moon.* Signed and dated 1948/9. Oil on canvas. 27½ in x 46 in (British Council, London)

48 *Portrait of Norman Douglas.* Signed and dated 1948 on back. Oil on metal. 10 in x 14 in (Sheffield City Art Galleries)

49 *The Shepherd.* 1949. Oil on board. 36 in x 40 in (Elisabeth Ayrton, Rockhampton)

50 *Good Friday.* Signed and dated 1949. Oil on canvas. 20 in x 16½ in (Bruton Gallery, Bruton)

51 *The White Cat.* Signed and dated 1949. Oil. Size unknown. (Private Collection, Italy)

52 *Olives and Vines.* Signed and dated 1947. Oil on canvas. 10 in x 14 in (Private Collection, Italy)

53 *Vines before Easter.* Signed and dated 1949. Oil on canvas. 24 in x 36 in (Elisabeth Ayrton, Rockhampton)

54 *The Passion of the Vine (VI).* Signed and dated 1949. Oil on canvas. 17 in x 37 in (Present whereabouts unknown)

55 *Oxen under the Vine.* Signed and dated 1950. Oil on board. 13½ in x 17½ in (Private Collection, England)

56 *Roman Window.* Signed and dated 1950. Oil on canvas. 48 in x 30 in (Swindon Museum and Art Gallery)

57 Michael Ayrton working in his Studio in All Souls Place, London, 1950, with *The Captive Seven* on the easel.

58 *The Elements as the Souces of Power* (1951). Gouache. Size unknown and presumed destroyed. This design was photographically enlarged to provide a mural in the interior of a ship's bows in the Festival of Britain, 1951, and the finished mural measured 68 ft x 120 ft.

59 *Girl on a Bed.* Signed and dated 1951/3. Oil on canvas. 20 in x 24 in (Present whereabouts unknown)

60 *Trio.* Signed and dated 30 November 1951. Brush, black ink and wash. 19 in x 25 in (Private Collection, U.S.A.).

61 *The Sisters* (1951). Oil on canvas. 30 in x 40 in (Present whereabouts unknown)

62 *The Sisters.* Signed and dated 1951. Brush, black ink and wash. 19 in x 25 in (Present whereabouts unknown).

63 *The Captive Seven.* Signed and dated 1950. Oil on canvas. 48 in x 72 in (D. & S. Hollis, London)

64 Study for *The Captive Seven.* Signed and dated 5/8 September 1950. Pencil. 7 in x 9 in (D. & S. Hollis, London)

65 *John Gielgud as Macbeth.* Signed and dated August 1942. Oil on board. 22½ in x 16¼ in (Bruton Gallery, Bruton)

66 *Figure in a Yellow Room.* (Portrait of Mme. Natalie Grandjean). Signed and dated 1951. Oil on canvas. 50 in x 28 in (D. & J. Hirschmann, Bristol)

67 *Portrait of John Minton.* Signed and dated 1941. Oil on panel. 12 in x 10¾ in (Private Collection, London)

68 *Portrait with Pears.* Signed and dated 1944. Oil on panel. 19¼ in x 23¼ in (Fay Pomerance, Sheffield)

69 *Portrait of Dylan Thomas.* 1947, signed. Brush and black ink and black wash on grey paper. 9 in x 6 in (John Arlott, Alresford)

70 *Portrait of William Walton.* Dated 24th May 1948. Pencil and black chalk. 9½ in x 6½ in (Boston Museum of Fine Arts)

71 *Portrait of Norman Douglas.* Signed and dated 1948. Brush, black ink and wash. 11 in x 8½ in (National Portrait Gallery, London)

72 *Portrait of Constant Lambert* (1951). Signed. Pen and black ink and black wash. 17 in x 12¼ in (Museum and Art Gallery, Salford)

73 *Portrait of William Walton, II.* Signed (1948). Oil. 24 in x 36 in (National Portrait Gallery, London)

74 *Portrait of Wyndham Lewis.* Signed and dated 1953. Oil on canvas. 30 in x 40 in (Private Collection, England)

75 *Portrait of Graham Sutherland.* Signed and dated 29th May 1967. Brush and black ink and black wash, over pencil. 16 in x 20 in (Gallery Huntly, Canberra)

76 *Portrait of William Golding.* Signed and dated 11th April 1965. Brush and black ink and black wash. 16 in x 20 in (Mrs William Golding, Bower Chalke)

77 *The Unforeseen Event (The Improbable Circumstance).* Signed and dated 3rd February 1952. Brush and black wash. 19 in x 25 in (Mrs Richard Law, London)

78 *Winter.* Signed and dated 1950. Oil on canvas. 20 in x 24 in. (Elisabeth Ayrton, Rockhampton)

79 *The Shepherds.* Signed and dated 1951. Oil on canvas. 30 in x 40 in (City Art Gallery, Manchester)

80 *The Shepherds.* Signed and dated 5th July 1951. Brush, black ink and wash. 20 in x 34 in (Present whereabouts unknown)

81 *The Shepherds: Large II.* Signed and dated 1951. Oil on canvas. 31½ in x 39½ in (Basil Wright, Henley-on-Thames)

82 *Study of a Shepherd.* Dated 21st March 1952. Black chalk over pencil. (Size and present whereabouts unknown)

83 *The Shepherd.* Signed and dated 1951. Lithograph.

84 *Green Still Life.* Signed and dated 1953. Oil on canvas. 19½ in x 28½ in (British Embassy, Rome)

85 *Still Life at Christmas.* Signed and dated 1953-54. Oil on canvas. 30 in x 40 in (Arts Council of Great Britain)

86 *Still Life—Carnations and Lemons on a Table.* Signed and dated 1953. Oil on canvas. 18½ in x 28 in (Basil Wright and Kassim Bin Said, Henley-on-Thames)

87 *Traveller's Joy.* Signed and dated 1953. Oil on canvas. 16 in x 18 in (John Crittall, Great Bardfield)

88 *Blind Girl with Sculpture.* 1952-53. Oil on canvas. 21½ in x 26 in (Dorothy Stein, Burford)

89 *Snowbird.* Signed and dated 1955. Oil on panel. 19½ in x 28 in (Philadelphia Museum of Art)

90 *Scavenger I,* 1954. Wax. 11¾ in x 16½ in by 6⅜ in. Remodelled in 1968 and cast as Scavenger II.

91 *Snowbird.* 1954-56. Bronze. 14 in high.

92 *Armed Head, Study for Bronze.* Signed and dated 2 August 1955. Monotype. 12 in x 9½ in (Elisabeth Ayrton, Rockhampton)

93 *Portrait of Henry Moore.* Signed and dated 18th January 1967. Pencil. 20 in x 30 in (Elisabeth Ayrton, Rockhampton)

94 *Talos, Armed Head—I.* 1954/56. Bronze. 20½ in x 13 in x 6 in

95 *Stylite,* 1958. Bronze. 42½ in x 9 in x 8½ in

96 *Figure on a Wall.* 1954. Bronze. 17 in x 11¾ in

97 *Still Life, Skull and Sculpture.* Signed and dated December 1954. Oil on panel. 40 in x 30 in (John W. Merriam, Philadelphia)

98 *Talos, Armed Head—II,* 1957. Bronze. 24 in x 14 in x 8 in

99 *Orpheus Seeking I,* 1958. Bronze. 21 in x 27 in x 10 in

100 *Orpheus Seeking II,* 1971. Bronze. 21½ in x 20 in x 12 in

101 *Sea Figure.* Signed and dated 12 August 1974. Black ink and wash over pencil. 16 in x 20 in (Private collection, U.S.A.)

102 *Falcon Watcher,* 1954/72. Bronze. 85 in x 26 in x 14¾ in

103 *Shepherd,* 1954. Bronze. 36 in x 9 in x 8 in

104 *Bather with Child II.* Signed and dated 3 January 1956. Pen and sepia wash, 24½ in x 19½ in (Christopher Allen, Bath)

105 *Bather with Child.* 1956. Bronze. 19 in x 4½ in x 4 in

106 Study for *Bather with Child III.* Signed and dated 1956. Pencil and oil wash on board. 24½ in x 19½ in (Present whereabouts unknown)

107 *Figures on a Jetty.* 1954/55. Bronze. 17½ in x 35 in x 10½ in

108 *The Swimmer.* 1952. Pen and ink and wash. 22 in x 32 in (Private Collection, U.S.A.)

109 *Bather with a Child.* 1956 (detail). Bronze. 19 in x 4½ in x 4 in

110 *Handstand.* 1956/57 (detail). Bronze and wrought iron. 39 in x 5 in x 7 in (Cartwright Hall, Bradford)

111 *Figure in Balance III.* 1956 (detail). Bronze and wrought iron. 26 in x 5 in x 5 in

112 *Double Scissored Figures.* Signed and dated 1956. Pen and sepia ink, sepia wash and chalk. 20 in x 30½ in (V.M. Elliott, Stoke-by-Clare)

113 *Acrobats II, Variant.* 1958 (detail). Bronze and wrought iron. 48½ in x 15 in x 11¼ in

114 *Treadmill.* 1974. Bronze and glass and x-ray film. 56¼ in x 15¾ in x 12 in (including base)

115 *Parade.* 1956. Signed and dated 1956. Pen and sepia ink and sepia wash. 23½ in x 20 in (Junior Common Room, Exeter College, Oxford)

116 *Catalan Cage Birds.* 1955. Signed and dated 1955. Oil. 60 in x 48 in (Walter S. Michel, New York)

117 *Cumae.* Signed and dated 12 May 1956. Brush and black ink and black wash. 15 in x 22 in (Elisabeth Ayrton, Rockhampton)

118 *The Acropolis at Cumae.* Signed and dated 11 May 1956. Pen and black ink and wash. 16 in x 20 in (Hollis Baker, Grand Rapids, Michigan)

119 *Mycenae—the Citadel.* 1957. Brush and sepia wash 14 in x 19 in (Trinity Hall, Cambridge)

120 *From Delphi towards Itea.* 1958. Signed. Brush and black ink and wash. 14 in x 20 in (Mrs Edward Warburg, New York)

121 *Dog at Phaestos.* 1957. Signed. Pen and black ink and black wash. 13¼ in x 19¾ in (Trevor Hickman, Wymondham)

122 *Figure in Landscape.* 1958. Bronze. 13 in x 25 in by 9½ in

123 *Perilous Place.* 1958. Wax and bone on wood. 16 in x 38 in (Elisabeth Ayrton, Rockhampton)

124 *Landscape of Cain.* 1958. Wax and bone on wood. 26 in x 42 in (Elisabeth Ayrton, Rockhampton)

125 *Midsummer.* 1958. Wax and bone on wood. 20 in x 25 in (Elisabeth Ayrton, Rockhampton)

126 *The Birth of Phanes II.* 1959. Wax and collage on wood. 30 in x 40 in (Elisabeth Ayrton, Rockhampton)

127 *The Birth of Phanes.* Signed and dated April 1959. Pen and ink and collage. 13 in x 20 in (Elisabeth Ayrton, Rockhampton)

128 *Icarus Falls.* Signed and dated 1959. Collage. 21½ in x 15½ in (Present whereabouts unknown)

129 *Icarus Falls.* 1959. Wax and bone. 60 in x 48 in (Heirs of Gavin Maxwell)

130 *Icarus Drowned.* 1960. Stucco Relief. 16 in x 36 in (National Gallery of New South Wales, Sydney)

131 *Slow Dance for the Nativity.* Signed and dated 1958. Oil on Board. 40½ in x 37 in (Elisabeth Ayrton, Rockhampton)

132 *Large Shore.* Signed and dated 1959/60. Acrylic and sand on panel. 30 in x 50 in (Lord and Lady Snow, London)

133 *The Terrace at Night.* Signed and dated 1959. Oil on panel. 25 in x 50 in (Ben Pomerance, Sheffield)

134 *Icarus Falls III.* Signed and dated 1959. Acrylic on canvas. 72 in x 40 in (Elisabeth Ayrton, Rockhampton)

135 *Large Cycladic I.* Signed and dated 1959/60. Acrylic and sand on canvas. 30 in x 50 in (Elisabeth Ayrton, Rockhampton)

136 *Corn Rhyton.* Signed and dated 1961. Acrylic on canvas. 25 in x 30 in (Bruton Gallery, Bruton)

137 *Icarus I.* 1959. Bronze. 19 in x 20 in x 8 in

138 *Daedalus Winged.* 1959/60. Bronze. 16 in x 17 in x 14 in (Metropolitan Borough of Sefton Libraries and Art Department)

139 *Icarus in Flight.* Signed and dated 1960. Plaster of Paris on wood. Painted. 16 in x 20½ in (The Arts Council of Great Britain)

140 *Icarus III Variant I.* 1960/72. Bronze. 67 in x 19 in x 24 in (Corporation of London: St. Paul's Cathedral in the background. A second cast is in the Royal Air Force Museum, Hendon)

141 *Icarus Pinwheel.* Signed and dated 3 October 1960. Charcoal. 20 in x 36 in (The Royal Air Force Museum, Hendon)

142 *Icarus Transformed III.* 1961. Bronze. 5½ in x 5 in x 18 in

143 *Icarus Transformed I.* 1961. Bronze. 8 in x 23 in 12 in (Tate Gallery, London)

144 *Icarus Winged.* Signed and dated 23 March 1960. Pen and black ink. 16 in x 10 in (John Metcalf, London)

145 *Figure under Stress.* Signed and dated 4 November 1961. Charcoal. 23½ in x 17½ in (T. G. Rosenthal, London)

146 *Blade Maze Figure.* 1965. Bronze. 19¾ in x 5 in

147 *Black Maze Figure.* Signed and dated 19 April 1965. White chalk on black paper. 22 in x 19 in (Nicholas Thompson, London)

148 *Emerging Maze Figure.* Signed and dated 14 February 1965. Pen and black ink wash with a strip of collage. 20 in x 16 in (Elisabeth Ayrton, Rockhampton)

149 *Translucent Maze.* 1965. Collage with charcoal on board. 30 in x 40 in (Elisabeth Ayrton, Rockhampton)

150 *Maze Maker II.* 1965. Bronze. 8½ in x 11 in x 14 in

151 *Maze Maker, Large Version.* 1965. Bronze. 34 in x 36 in x 24 in

152 *Maze Maker II (Copper Seam).* Signed and dated 1966. Acrylic and collage on board with gold leaf. 60 in x 40 in (Elisabeth Ayrton, Rockhampton)

153 *Daedalus, the Wing Maker.* Dated 1960. Bronze. 18 in x 22 in x 15 in

154 *Daedalus at Cumae.* 1961. Bronze. 21 in x 11 in x 7½ in.

155 *Sun Maze.* 1970. Bronze. 16½ in x 12 in x 4 in

156 *Daedalus/Icarus, Matrix I.* 1968. Bronze. 17¼ in x 8¼ in x 4½ in

157 *Daedalus/Icarus, Matrix II,* 1968. Bronze. 18 in x 9¼ in x 10 in

158 *Figure with a Skein II.* 1963/72. Bronze. 35 in x 7¼ in x 13 in

159 *Crouched Minotaur.* 1963. Bronze. 26 in x 34 in x 21 in

160 *The Evolution of the Minotaur.* 1964 (detail). Bronze. 67 in x 29 in x 18 in

161 *Restless Minotaur.* 1964. Bronze. 11½ in x 6 in x 6½ in

162 *Minotaur in Jeopardy.* 1964. Bronze. 11½ in x 6 in x 6½ in

163 *Minotaur Enraged.* Signed and dated 24 December 1963 Charcoal. 25 in x 18½ in (Elisabeth Ayrton, Rockhampton)

164 *Minotaur Erect.* 1975. Bronze. 26 in x 7 in x 12 in

165 *Daedalus/Icarus Matrix.* Large Version. 1968/69. Bronze. (Sue Erpf van de Bovenkamp from Estate of Armond G. Erpf)

166 *Arkville Minotaur.* 1968/69. Bronze. 55 in x 42 in x 36 in (Sue Erpf van de Bovenkamp from Estate of Armond G. Erpf. A second cast is in the collection of the Corporation of London and may be seen in Postman's Park near Guildhall)

167 *Minotaur Surprised.* Signed and dated 10 October 1975. Pen and black ink and black wash. 20 in x 16 in (Elisabeth Ayrton, Rockhampton)

168 *Minotaur Alarmed.* Signed and dated 23 July 1963. Brown and black ink and wash. 20 in x 16 in (George Steiner, Cambridge)

169 *Minotaur Waking.* Bronze. (Southampton Art Gallery: City of Cork)

170 *Minotaur Risen.* 1971. Bronze. 8¾ in x 5 in x 5½ in

171 *Arkville Minotaur* (detail). 1968/69.

172 *The Arkville Maze* (Sue Erpf van de Bovenkamp)

173 *Arkville Maze—Maquette.* 1968. Fibreglass and bronze. 27½ in x 37 in x 2½ in

174 *Suite of Ten Minotaur Etchings.* 1971. Embryo. Consecrated. As Calf. As Yearling. Rising. Risen. Full Grown. Pent. Revealed. Alone. (British Museum, London)

175 *Oracle. Version I.* 1962. Bronze. 17 in x 8½ in x 13 in

176 *Pythia.* 1963. Bronze. 13 in x 8 in x 8 in

177 *Tripod Oracle.* 1962. Bronze. 17 in x 3 in x 5 in

178 *Mantic Figure.* 1963. Bronze. 15 in x 5 in x 4¾ in

179 Study for *Tripod Oracle.* Signed and dated 25 November 1962. Brush and black ink and black wash. 16 in x 10 in (Elisabeth Ayrton, Rockhampton)

180 *Siren.* 1963. Bronze. 9½ in x 4 in x 2½ in

181 *Mantic Figure II.* Signed and dated 25 November 1962. Pen and black ink. 10 in x 9 in (Elisabeth Ayrton, Rockhampton)

182 *The Oracle as Pythia.* Signed and dated 16 February 1964. Pen and black ink. 20 in x 25 in (Junior Common Room, Exeter College, Oxford)

183 *Oracle—Responding Head I.* Dated 4 August 1963. Brush and black ink and wash. 16 in x 20 in (Winston Graham, Buxted)

184 Study for *Talos.* Signed and dated September 1963. Black ink and red wash. 22 in x 18 in (Keatley Trust, Royston)

185 *Demeter and Kore.* Dated 16 July 1965. Charcoal drawing on canvas. 30 in x 25 in (Private Collection)

186 *Seasons of Demeter: I Spring.* Dated 4 February 1965. Pencil. 20 in x 16 in (Grosvenor Gallery, London)

187 *Seasons of Demeter: IV Winter.* Dated 4 May 1965. Pencil. 20 in x 16 in (Grosvenor Gallery, London)

188 *Seasons of Demeter: II Summer.* Signed and dated 29 January 1965. Pencil. 16 in x 20 in (Elisabeth Ayrton, Rockhampton)

189 *Demeter Pregnant.* 1966. Bronze. 31 in x 13 in x 14 in (Reading Museum and Art Gallery)

190 *Winter Demeter.* 1967. Bronze. 11½ in x 6 in x 7½ in

191 *Seasons of Demeter: III Autumn.* Signed and dated 4 May 1965. Pencil. 16 in x 20 in (Elisabeth Ayrton, Rockhampton)

192 *Pupate Figure.* Signed and dated 22 November 1965. White chalk on black paper. 16 in x 20 in (Michael's Restaurant, Bristol)

193 *End Maze II.* 1970. Bronze. 20½ in x 22 in x 4½ in

194 *Return.* 1969. Bronze. 5 in x 9 in x 3½ in

195 *Carapace.* 1972. Bronze. 5½ in x 8½ in x 7 in (Royal Albert Memorial Museum, Exeter)

196 Study for *End Maze, Drawing III.* Signed and dated 1970. Pen and diluted black ink. 16 in x 20 in (Elisabeth Ayrton, Rockhampton)

197 *Berlioz in Rome, 1831.* 1947. Pencil. 10 in x 14 in (Elisabeth Ayrton, Rockhampton)

198 *Berlioz in 1852.* 1968. Acrylic. 40 in x 30 in (Elisabeth Ayrton, Rockhampton)

199 *Sleeping Figures.* 1967. Bronze. 2 in x 12½ in x 9 in

200 *Emerging Figure II.* Signed and dated 24 January 1966. Pencil. 20 in x 16 in (Elisabeth Ayrton, Rockhampton)

201 *Afternoon.* 1973. Bronze. 7¼ in x 12 in x 20 in

202 *Contained Maze,* Signed and dated April 1966. Acrylic and collage on canvas. 40 in x 60 in (Elisabeth Ayrton, Rockhampton)

203 *Triptych II Emerging Figure.* 1967. Bronze. 14¾ in x 32 in x 12 in

204 *Through the Blade II.* 1969. Bronze. 11 in x 6 in x 4 in

205 *Penetrator II.* 1969. Bronze. 17½ in x 22½ in x 14 in (Elisabeth Ayrton, Rockhampton)

206 *Penetrator.* Signed and dated 9 February 1970. Pencil. 16 in x 20 in (Elisabeth Ayrton, Rockhampton)

207 *Troy Maze.* 1972. Bronze. 29 in x 48 in x 18 in

208 *Troy I.* Signed and dated 16 August 1972. Pen and black ink over pencil. 10 in x 14 in (Elisabeth Ayrton, Rockhampton)

209 *Mirror Maze.* 1966. Bronze and threads. 14 in x 24 in x 18½ in

210 *Mirrored Figure I.* 1970. Bronze and perspex. 19 in x 8 in x 5¼ in

211 *Dioskouri* 1970. Bronze and perspex. 19¼ in x 14½ in x 8½ in

212 *The Discovery of Nautilus.* 1969. Bronze and perspex. 15½ in x 22¼ in x 15½ in

213 *Mirror Twins.* 1973. Bronze and perspex. 16 in x 9 in x 8 in

214 Study for *The Discovery of Nautilus II.* Signed and dated 1970. Pen and wash and black ink. 16 in x 20 in (Private Collection, England)

215 *Reflective Heads II.* 1971. Bronze. 16 in x 10 in x 7 in (City Art Gallery, Southampton)

216 *Brain Maze (Drawing III).* Dated 19th September 1972. Pen and diluted ink. 10 in x 14 in (Elisabeth Ayrton, Rockhampton)

217 *Reflex II.* 1969. Bronze and perspex. 13 in x 18½ in x 10 in

218 *Brain Maze IV.* 1973. Pen and black ink and wash. 14 in x 10 in (Elisabeth Ayrton, Rockhampton)

219 *Brain Maze.* 1972 .Bronze and perspex. 11¼ in x 8¾ in x 10¾ in

220 Study for *Reflective Head.* Dated 27 October 1971. Pen and wash and black ink over pencil. 16 in x 20 in (Kresge Company, Troy, Michigan)

221 Study for *Reflective Head.* Dated 28 October 1971. Pen and brush and black ink. 10 in x 14 in (Kresge Company, Troy, Michigan)

222 *Reflective Heads.* 1971/72. Bronze and plate glass. (Kresge Company Headquarters, Troy, Michigan)

223 *Translucent Heads II.* Signed and dated 1970. Black ink and wash. 10 in x 14 in (Detroit Institute of Arts)

224 Study for *Contained Heads I.* Signed and dated 24 March 1970. Brown and black ink and wash. 16 in x 20 in (Timothy Easton, Debenham)

225 *Maze Head.* Signed and dated 24 April 1965. White chalk on black paper. 16 in x 20 in (Jorge Lewinski, London)

226 *Maze Head.* 1965. Bronze. 10 in x 9 in x 11 in

227 *Constructed Heads.* Signed and dated 15 March 1972. Pencil. 20 in x 16 in (Elisabeth Ayrton, Rockhampton)

228 *Tricephalic.* 1972. Bronze. 24¾ in x 9 in x 9 in

229 *Tricephalic.* Signed and dated 5 November 1972. Collage with pen, sepia and black ink. 20 in x 30 in (Private Collection, U.S.A.)

230 *Maze Music, Drawing II.* Signed and dated 30 October 1972. Sepia ink and red chalk. 16 in x 20 in (Private Collection, England)

231 *Maze Music.* 1972. Bronze and threads. 42 in x 35 in x 11 in (Portsmouth City Art Gallery)

232 *Red Thread.* 1972. Bronze and threads. 7¼ in x 12¼ in 3½ in

233 Study for *Web.* Signed and dated 21 April 1973. Pen and sepia ink and red chalk. 18 in x 14 in (Private Collection, U.S.A.)

234 *Web.* 1972. (Detail). Bronze, steel and threads. 69½ in x 13½ in x 10 in

235 *Laocoön Maze Figure III.* Dated 25 April 1972. Pen and sepia ink. 14 in x 20 in (Elisabeth Ayrton, Rockhampton)

236 *Laocoön Maze Figure II.* 1972. Bronze. 9½ in x 8¾ in x 3½ in

237 *Alberti.* 1974. Bronze. 11½ in x 7 in x 5½ in

238 *Transformation.* Signed and dated 9 July 1974. Brush and black ink and black wash over pencil. 14¾ in x 19 in (Elisabeth Ayrton, Rockhampton)

239 *Eclipse.* Signed and dated October 1974. Pen and brush and black ink over collage. 20½ in x 24½ in (Private Collection, U.S.A.)

240 *Heavy Figure I.* Signed and dated 2 November 1974. Pencil, pen and black ink and wash. 16 in x 20 in (Elisabeth Ayrton, Rockhampton)

241 *African Buffaloes.* Signed and dated 1975. Sepia and black ink. 16 in x 20 in (Prue Hopkins, Rockhampton)

242 *Warthogs.* Signed and dated 12 February 1975. Black ink and wash. 10 in x 14 in (Prue Hopkins, Rockhampton)

243 *Baboons.* 1975. Etching.

244 *Tsavo East.* Signed and dated 24 March 1973. Pen and brown ink and brown wash, with pastels, on grey paper. 20 in x 16 in (Prue Hopkins, Rockhampton)

245 *Suite of Archilochos Etchings.* 1975. Paros. Epode. Spears. Eclipse. Lykambes and his Daughters. Salmydessos. Archilochos Alone. Impact.

246 *Point of Departure.* 1970. Bronze. 14¾ in x 8¾ in x 9¾ in

For reasons of limitation of space only the names of public collections are given in relation to bronzes; these are cast in limited editions.

Photographic Acknowledgements

Birmingham Museums and Art Gallery: 13 14 15 16 17 19 20 24 25 27 35 37 38 39 40 55 67 68 75 244

Brompton Studio, London: 130

A.C. Cooper, London: 1 2 3 5 6 8 9 18 21 22 23 26 28 29 30 31 32 33 34 36 42 43 44 45 47 48 49 50 51 52 53 54 56 59 60 61 62 64 66 69 70 71 72 73 74 77 78 79 80 81 82 84 86 87 88 89 96 97 98 104 106 107 108 109 110 111 112 115 118 126 127 131 132 133 135 141 143 145 147 154 186 187 202

G. Faiers, Sible Hedingham: 90 94 99 100 101 114 117 119 120 148 150 152 163 164 167 173 179 181 184 193 194 196 198 199 201 206 207 208 211 213 214 216 217 218 220 221 223 224 227 228 229 230 233 234 235 241 242

R.B. Fleming, London: 63

J.R. Freeman, London: 85 139

B. Korab, Troy, Michigan: 222

M. Le Marchant, Bruton: 65 91 95 105 113 122 128 140 157 160 162 166 169 170 195 204 212 219 231 232

J.S. Lewinski, London: 76 93 146 149 156 159 168 185 188 189 190 191 192 200 203 225 226

W. Eaden Lilley, Cambridge: 153

The Manor Studio, Southall: 92 116

Henry Morgan, Yeovil: 102 210 215 236 237 238 239 240

Photo Studios, London: 176

The Redfern Gallery, London: 46

Royal College of Art: Frontispiece

Stearn and Sons, Cambridge: 136 137 161 175 177 178 180 183

John Vickers, London: 4 103 123 124 125 129

Victoria and Albert Museum, London: 7 10 11 12 41

John Webb, London: 155 197 209